What teachers
need to know about
Students with
disabilities

What teachers
need to know about
Students with
disabilities

PETER WESTWOOD

ACER Press

First published 2009
by ACER Press, an imprint of
Australian Council for Educational Research Ltd
19 Prospect Hill Road, Camberwell
Victoria, 3124, Australia

www.acerpress.com.au
sales@acer.edu.au

Reprinted 2010, 2012

Edited by Elisa Webb
Cover and text design by Mason Design
Typeset by Mason Design
Printed in Australia by BPA Print Group

National Library of Australia Cataloguing-in-Publication data:

Author: Westwood, Peter S. (Peter Stuart), 1936–
Title: What teachers need to know about students with disabilities /
            Peter Westwood.
ISBN: 9780864318695 (pbk.)
Series: What teachers need to know about
Notes: Includes index.
            Bibliography.
Subjects: Students with disabilities—Education.
            Teachers of children with disabilities—Training of.
            Learning disabilities.
            Reading disability.
Dewey Number: 371.91

Contents

Preface

The advent of the inclusive education movement has seen mainstream schools and mainstream teachers involved to a much greater extent than ever before in accommodating students with a wide range of disabilities. For this reason, all mainstream teachers need to acquire a good working knowledge of the nature of disabilities and their potential effects on learning and development. In particular, teachers need to be aware of research-based methods that have proved most effective in meeting the educational needs of these students. This book aims to provide teachers with this essential information in a form that is easy to read and understand.

There is much in this book that is also highly relevant for teachers and paraprofessionals working in special schools and centres. While pro-inclusion organisations such as the Centre for Studies in Inclusive Education (UK) and the Alliance for Inclusive Education (UK) espouse a view that *every child*, regardless of ability or disability, has the right to education in his or her local mainstream school, the reality is that such an arrangement is neither feasible nor desirable. It will be evident from some of the data presented in this book that the type of intensive and structured teaching and management required by some students with severe and complex cognitive and behavioural difficulties are very different from the teaching methods commonly used with more independent learners. To attempt to provide such specialised teaching for certain individuals within the context of a regular mixed-ability classroom makes absolutely no sense. For some students with severe disabilities, appropriate education can be provided most effectively in a special school – and for these students a special school is the least restrictive learning environment.

Very many students with milder forms of disability have obviously benefited, and will continue to benefit, from inclusive education; but as Runswick-Cole (2008, p. 173) states: 'The process of inclusive education continues to be fragile'. After well over a decade of inclusive education policies, the number of students attending special schools for intellectually disabled students (for example) has remained fairly stable (Norwich, 2008). This is surely evidence not that policies are failing, but that such specialised provision is meeting a very real need. Pursuing policies of '*full* inclusion' and 'zero rejection' for ideological reasons is simply not in the best interests of these most severely disabled students. As Topping and Maloney (2005) observe, providing all students with equal opportunities does not mean treating everyone equally, because that simply reinforces differences. Rather, notions of social justice and equality of opportunity imply that often we need to treat different students differently so that they will have equal opportunities to maximise their potential. Students with severe and complex disabilities represent a major case in point.

This book is not intended to be an exhaustive list of all disabilities and disorders, nor to provide comprehensive medical diagnoses. The aim of this book is to give teachers an overview of the disabilities that they are more likely to encounter in their classrooms, and how these disabilities affect students' learning and development. Teachers should also be aware that advances continue to be made in medical treatment and therapy. The links at the end of each chapter, together with the main list of references, can be used by readers requiring more detailed information.

PETER WESTWOOD

RESOURCES   **www.acer.edu.au/need2know**

Readers may access the online resources mentioned throughout this book through direct links at www.acer.edu.au/need2know

Intellectual disability

KEY ISSUES

▸ Students with intellectual disability comprise a significant and diverse group of individuals with special educational needs.

▸ Intellectual disability has many causes. Knowing the cause in individual cases does not necessarily help with selection of teaching methods.

▸ Teaching methods should be selected to match the observed learning characteristics, strengths, weaknesses and needs of the student.

▸ Early intervention is vital for children with intellectual disability.

It is estimated that individuals with intellectual disability comprise some 3 per cent of the general population (Prater, 2007). This may not seem a large figure, but students with intellectual disability actually comprise a significant group of individuals with special educational needs in the school system, far outnumbering those with serious impairments of vision or hearing or with physical disabilities.

General characteristics of students with intellectual disability

The most obvious characteristic of students with intellectual disability is that they experience great difficulty learning almost everything that other students learn with comparative ease. They seem to lack the ability to think quickly, reason deeply, remember easily, plan ahead and adapt rapidly to new situations. Students with this disability usually appear much less mature than their age peers, often exhibiting a range of behaviours and responses that are typical of younger children.

Regardless of these obvious weaknesses, students with intellectual disability are still capable of much learning and can acquire many new behaviours, particularly when effective teaching methods are used. Many of these students have surprised their parents and their teachers by their positive progress once they are in a supportive school situation. It is important to hold high, rather than low, expectations for their progress. All students with intellectual disability have areas of relative strength, and teachers and parents need to identify these strengths in order to build upon them. Effective methods and teaching principles to achieve optimum progress are described later in this chapter.

As individuals, students with intellectual disability vary widely in their levels of functioning, their degree of self-management, their strengths and weaknesses, their personalities, their ability to cope with schoolwork, their response to instruction, and the extent to which they have additional physical or sensory impairments, medical conditions and emotional or behavioural problems (Hodapp & Dykens, 2003; Myrbakk & von Tetzchner, 2008). Some of these differences are innate and are caused by the student's particular syndrome of intellectual disability. Other variations stem from different levels of care, stimulation and support they have received within their family and school environments.

The two overriding features of intellectual disability are:

▶ significant limitations in *cognitive functioning*, together with
▶ major difficulties acquiring *adaptive behaviours* (AAMR, 2002).

The term 'cognitive functioning' refers to the ability to acquire knowledge, think and reason, communicate, and learn from one's own actions and from instruction. The term 'adaptive behaviour' refers to the functional abilities that are necessary for independence in everyday life such as self-care, social competence and self-management. Hallahan et al. (2009) suggest that effective adaptive behaviour relies upon a combination of *social intelligence* (understanding social situations and interacting appropriately with others) together with *practical intelligence* (ability to solve everyday problems, acquire self-help skills and function independently). Individuals with intellectual disability are frequently lacking in both the social and the practical aspects.

Identifying intellectual disability

Traditionally, cognitive functioning has been assessed by the use of appropriate intelligence tests. Individuals with measured IQs below 70 have been regarded as intellectually disabled, particularly if they also display major difficulties in acquiring everyday skills and self-management. In recent years however, adaptive behaviour has been given greater importance than IQ in the identification of intellectual disability because deficiencies in adaptive behaviour indicate more clearly the amount and intensity of support an individual may require (Batshaw et al., 2007; Cohen & Spenciner, 2005).

Causes of intellectual disability

The causes of intellectual disability often originate at conception or birth, and more than 750 separate factors have been discovered that can potentially cause intellectual disability. When it comes to assessing individual cases however, unless the individual has a recognised syndrome or genetic disorder, it is often impossible to identify with certainty the exact cause of his or her impairment (Heward, 2009). This is particularly the situation with individuals who have a mild degree of disability and no additional handicaps.

Causal factors are usually grouped into two categories, *biological* (organic) and *psycho-social* (environmental) (Brown, et al., 2007). Research evidence suggests that only 25 per cent of intellectual disability is due to biological causes, with the remaining 75 per cent due mainly to psycho-social disadvantage or to unknown causes. Psycho-social disadvantage is a term implying that a child's intellectual development may have been depressed in the early formative years by being reared in an adverse and impoverished environment (Henley et al., 2009). Intensive early intervention programs for such children have often resulted in significant cognitive gains, thus supporting a view that their intellectual impairment was due, at least in part, to negative environmental influences (Weiten, 2001). In many cases, intellectual disability may be caused by a combination of poor genetic endowment interacting with extremely adverse environmental influences (Blackbourn et al., 2004). The terms 'socio-cultural disadvantage' and 'cultural–familial deprivation' have also been applied to these environmental causal influences.

Biological causes of intellectual disability include:

- chromosomal anomalies (e.g., Down syndrome; Fragile X syndrome; Rett syndrome) (Gordon, 2007; Roizen, 2007)
- inborn errors of metabolism resulting in enzyme deficiencies and structural damage to the brain (e.g., phenylketonuria) (Gordon, 2007)
- prenatal adverse influences (e.g., diseases of the mother during pregnancy; mother's exposure to alcohol, drugs or other toxins during pregnancy; severe maternal malnutrition) (Bell, 2007; Davidson & Myers, 2007)
- perinatal hazards (e.g., difficult birth; brain damage due to anoxia; prematurity; very low birth weight) (Batshaw et al., 2007)
- neonatal factors (e.g., diseases in infancy affecting the brain; meningitis or encephalitis; traumatic head injury; accidental ingestion of lead or other forms of poisoning; severe malnutrition) (Batshaw et al., 2007).

Psycho-social causes of intellectual disability include:

- poverty
- malnutrition
- chronic abuse
- physical and emotional neglect
- parents of low cognitive ability and/or with psychiatric disorders
- lack of stimulation
- lack of opportunity.

Santrock (2006) suggests that most individuals with biological causes of intellectual disability tend to have IQs below 50, while those with psycho-social disadvantage tend to have IQs from 50 to 70. Those individuals with IQs above 50 tend to be the most responsive to instruction and are most likely to benefit from full-time or part-time inclusion in mainstream programs.

Children with different forms of intellectual disability may differ in their behaviour, strengths, weaknesses, response patterns and motivation (Hodapp & Dykens, 2003). Teachers need to be concerned with identifying the knowledge, skills and adaptive behaviours that a student has already acquired, and determining what he or she needs to learn next in order to set relevant goals for that student's educational program (Cronin & Patton, 1993; Snell & Brown, 2006).

Degrees of disability

It is common to discuss intellectual disability in terms of degree of relative severity, usually defined as either mild, moderate or severe to profound (Cohen & Spenciner, 2005). The percentage of individuals in each category reduces significantly as one moves down the scale from mild (85 per cent of cases), moderate (10 per cent), and severe to profound (5 per cent). The amount of support required for learning and for everyday living increases steadily as the degree of severity increases. As indicated above, the causes of most severe and profound disability are more likely to be organic rather than environmental; and in such cases the intellectual disability is likely to be accompanied by additional physical, sensory, psychiatric or behavioural impairments (Batshaw et al., 2007; Felce et al., 2009; Myrbakk & von Tetzchner, 2008).

Mild disability

Most students with mild intellectual disability can be educated effectively in inclusive classrooms, particularly in the primary school years. However, they do require some modification to mainstream curriculum in academic subjects, and usually benefit from intensive remedial teaching in language, literacy and numeracy areas. They may also require help in the domain of social relationships since many of these children have difficulty making friends. Some may also exhibit behaviour problems (Embregts et al., 2009).

The vast majority of individuals with intellectual disability are located in this *mild* category, with IQs in the range 55 to 70. These students are very little different in most respects from those with IQs in the range 70 to 85 who in the past have been referred to as 'slow learners' or 'low achievers'. Most students with mild disability appear physically and behaviourally very similar to children without disabilities. Outside the school context, the majority of these students are considered completely normal, their problems only becoming apparent when they experience learning difficulties within the school curriculum. Their lack of success can often lead to loss of self-esteem, loss of confidence and reduced motivation to learn. They tend to blame their failures on their own lack of ability and this reduces their inclination to persevere when schoolwork becomes challenging. Most mildly disabled individuals manage to live fully independent lives (although not necessarily highly successful lives) after leaving school. Weiten (2001) observes that most students in this category reach sixth-grade standard of scholastic achievement by the time they are in their late teens.

The underlying cause of mild intellectual disability in individual cases is usually unknown (Heward, 2009). In a few instances the cause may be organic (e.g., Fragile X syndrome; Down syndrome; Foetal Alcohol Syndrome[1]) (Dew-Hughes, 2004; Nulman et al., 2007; Wishart, 2005), but in the majority of cases the causes are mainly environmental. Students from lower socio-economic status families, from dysfunctional families, from minority groups and from depressed environments are over-represented in this mild disability category (Cartledge et al., 2009; Hodapp & Dykens, 2003).

Moderate disability

Students whose degree of intellectual disability is in the moderate range (IQs between 40 and 54) experience significant learning difficulties. They usually require full-time special education and only a few are accommodated successfully in mainstream classes, usually only in the early school years. With intensive and effective teaching they can reach an achievement level somewhere between second and fourth grade by the time they are in their late teens (Santrock, 2006).

These students may display significant deficits in adaptive behaviour, so their educational program must focus on teaching skills for self-management, social competence and communication. Ongoing behaviour management programs are also required for a few of these students whose lack of self-control can result in very challenging behaviour.

Severe to profound disability

Measuring the IQ of individuals with severe or profound intellectually disability is extremely difficult and somewhat inaccurate. However, most experts assign the IQ range 25 to 39 for severe disability and below 25 for profound disability. This degree of cognitive impairment causes extreme difficulties with learning, communication and self-care. Students functioning at this level need constant management and support, often throughout their lives. Some of these students lack both mobility

---

1   Information on these syndromes is provided in Chapter 2.

and speech, so physical therapy is necessary and alternative methods of communication must be established (e.g., sign language, or picture and symbol communication) (Bondy & Frost, 2003). Intensive training, using mainly direct teaching and the consistent application of behavioural principles (task analysis and reinforcement) is required to teach these students the most basic of self-help skills such as eating, toileting, washing and dressing. Some individuals with profound intellectual and multiple disabilities (PIMD) never manage to accomplish these daily living skills unaided and a few remain relatively unresponsive to training (Weiten, 2001).

It is quite common to find that students with severe and profound disability have little control over their emotions or behaviour (Batshaw et al., 2007). They may exhibit aggression, self-injury and stereotypic obsessive-compulsive behaviours (Cooper et al., 2009). Behaviour modification techniques are usually required to reduce or eliminate these detrimental behaviours (see Chapter 7).

In special schools, methods such as *intensive interaction* (Caldwell, 2006; Firth, 2009) and *Snoezelen* (Hutchinson & Kewin, 1994) may be used with the most severely disabled students to stimulate responses. Intensive interaction is similar to the natural approach used instinctively by parents when responding to a very young child's actions – for example by smiling, hugging, and praising. Anything that the child does spontaneously leads the adult to respond actively. This strengthens the child's motivation to initiate actions and to communicate. *Snoezelen* is a therapeutic approach developed first in Holland but now used in many countries. It provides both sensory stimulation and relaxation therapy for the most severely handicapped individuals. More research is needed to evaluate the extent to which these approaches really help these students to learn.

Learning characteristics of individuals with intellectual disability

The learning characteristics of persons with intellectual disability have been the focus of a great deal of research attention for more than a century. Different forms of intellectual disability often result in different patterns of behaviour and different responses to education and training. However, researchers have discovered that almost all individuals with this disability manifest many of the same basic underlying problems in learning (Hodapp & Dykens, 2003). These problems stem from deficiencies in selective attention, speed of information processing, memory, reasoning, concept development, language and transfer of learning (Prater, 2007). In addition, students with intellectual disability tend not to develop effective approaches to instructional tasks because they lack metacognitive awareness that would enable them to reflect upon the results of their own actions (Drew & Hardman, 2007). Some experts argue that not knowing how to learn effectively is actually their greatest impairment (Hallahan et al., 2009; Pressley & McCormick, 1995). For example, these students rarely regulate their approach to learning activities by using 'self-talk', instead responding impulsively and producing high error rates. Effective teaching for these students must include explicit instruction in appropriate procedures for approaching each new task.

Attention and information processing

One of the most noticeable characteristics of students with intellectual disability is their poor level of concentration in a formal learning situation. Specifically, they have great difficulty focusing attention selectively, sustaining attention, dividing attention as necessary between different aspects of a task and processing sequential information. This phenomenon has been recognised for many years (e.g., Zeamon & House, 1963). In addition, these students have a tendency to be easily distracted by events occurring around them in the classroom. Poor attention significantly reduces the ability to take in information and to learn from observation and imitation. Panda (2001) suggests that attention problems are actually more significant than any other cognitive impairment in accounting for the learning difficulties of these students. Distractibility and limited attention span can cause significant problems for a child with intellectual disability when integrated into a mainstream classroom.

Teachers need to be fully aware of this pervasive attention difficulty and take it into account when working with and instructing intellectually disabled students. It is a well-established fact that the first step in effective teaching is gaining and holding the learner's full attention. This usually requires capturing and maintaining interest or curiosity – for example, by using mainly visual and concrete materials and requiring active modes of responding (Downing & Eichinger, 2008). Heward (2009) suggests that students are more likely to sustain attention to tasks if they are experiencing success. Poor attending behaviour during initial teaching also highlights the need for frequent repetition, review and additional practice.

Partly as a result of poor attending behaviour (but also existing as a separate weakness in some cases) students with intellectual disability take much longer to process and respond to information (Bennett, 1999; Pressley & McCormick, 1995). Vaughn, Bos and Schumm (2007) remind us that students with intellectual disability learn at a much slower rate and are particularly challenged and confused by complex, abstract or multi-tiered tasks. These learning characteristics suggest a need to break complex tasks down into many smaller steps, allowing sufficient time for students to understand instructions and to give their responses (Downing & Eichinger, 2008).

Memory

Many students with intellectual disability exhibit great difficulty storing information in long-term memory (Pickering & Gathercole, 2004; Prater, 2007). In most cases, this weakness is closely associated with their failure to attend, as described above. But these students appear also to have no effective strategy for memorising important information, or indeed, for recognising when information needs to be remembered. Sternberg (2003) views such memory deficits in people with intellectual disability as being due to underdevelopment of important metacognitive processes that would help them link new information with what they already know.

Whatever the underlying cause, the effect of limitation in memory is that it takes much longer, and many more repetitions, to get information finally stored. Heward (2009) reminds us that earlier research work on memory seems to indicate that when an intellectually disabled person has finally understood information and committed it to long term memory, he or she can retain and recall that information over time as proficiently as any other individual.

Thinking, reasoning and concept development

Thinking and reasoning are complex cognitive processes involving among other things: the appropriate use of stored information and previously acquired concepts; the ability to relate what is observed to what is already known; questioning; recognising cause and effect connections; reflecting;

forming and testing hypotheses; drawing conclusions and developing opinions and beliefs (Eggen & Kauchak, 2007). Individuals with intellectual disability clearly have major weaknesses in thinking and reasoning, and this makes most forms of problem-solving or advance planning very difficult for them. This should not surprise us because 'intelligence' is sometimes defined simply as 'the ability to think and to reason'. Impairment in intellectual functioning almost automatically implies difficulty in executing these key cognitive processes.

Specific areas of the brain are thought to be involved in reasoning (Weiten, 2001), so any organic damage or dysfunction associated with those areas, as occurs in some forms of intellectual disability, may well impair the ability to reason. Pressley and McCormick (1995, p. 3) state:

> Good information processing requires a generally healthy, normally functioning brain. ... Much of thinking also depends on short-term memory, variously referred to as consciousness, working memory, or short-term storage. Short-term memory depends on an intact brain.

The work of Piaget (e.g., Piaget, 1952) revealed that the depth, quality and sophistication of children's thinking changes significantly during the years between birth and late adolescence. Their cognitive development passes through a number of stages from pre-conceptual, through a concrete operational stage, to the final stage where formal and abstract reasoning is possible. Individuals with intellectual disability pass through these same stages of development but at a much slower rate, and the majority may never reach the upper stage of abstract (formal) reasoning. Most students with mild or moderate disability remain firmly at the concrete operational stage – they understand and remember only those things and situations that they can directly experience and that they can link with what they already know. Students with severe and complex disabilities may be operating cognitively at even lower (pre-operational) levels and are unable to use inductive or deductive reasoning (Eggen & Kauchak, 2007).

A large part of thinking, reasoning and learning involves the acquisition and use of concepts. A concept can be defined as a mental representation that embodies all the essential features of an object, a situation or an idea. Concept formation is the means by which we mentally organise our environment into meaningful units of information that we can use for future reference. Thinking relies heavily on the ability to draw upon concepts that have been developed and stored in long-term memory. The process of forming concepts consists of classifying and linking items of information together because of common properties they possess. For example, our concept of 'house' embodies our knowledge of the outside appearance of many different styles and forms of housing, awareness of the typical internal arrangement of rooms and their functions, and the normal purposes for which a house is used. We also recognise this concept as falling within a larger concept embracing the notion of 'buildings'. Conceptual knowledge can be thought of as a connected web of information that facilitates thinking. Prater (2007) observes that students with intellectual disability are much less skilled than others in linking together and consolidating separate items of information; so for them, concept development is much slower. Their paucity of concepts places a limit on their ability to think and reason.

Language skills

Language serves a multitude of purposes in learning, and in personal and social development. For example, language enables a person to make his or her needs and ideas known to others, to ask or answer questions in order to obtain or give information, to interact socially with others and to function effectively in school. Even more importantly, language is a key facilitator of cognitive development, because words represent much of the raw material we use when thinking and reasoning. We assimilate and store concepts much more effectively if they have a representation in words. When an individual lacks adequate language skills, he or she has much greater difficulty learning – both in and out of school – and in developing socially.

One of the most obvious characteristics of students with intellectual disability, particularly those with moderate and severe to profound disabilities, is the very slow rate at which they acquire communication skills (Kuder, 2008). Indeed, many individuals with severe and multiple disability never develop speech and need to be taught to use alternative methods of communication (Heller & Bigge, 2005). Many students with intellectual disability require speech therapy (Hegde, 2008), but their improvement in this area can be very slow indeed. This poor response is because students receiving therapy may not appreciate the need for it and may therefore lack motivation to practise what is taught. In addition, there is lack of transfer of what is taught in a clinical setting to everyday situations.

The development of verbal and non-verbal communication skills is given very high priority in early intervention programs and in special school curricula (Beirne-Smith et al., 2006). One of the benefits that should occur when a student with intellectual disability is included in mainstream schooling is increased exposure to good models of language and an increased need to communicate.

Generalisation and transfer of learning

Hardman et al. (2005) indicate that intelligent behaviour is associated with the ability to apply what is learned to new situations. In the context of school learning, generalisation and transfer occur when something learned at one time and in one place is applied without prompting in another setting (Eggen & Kauchak, 2007). For students with intellectual disability, generalisation is often a major hurdle because their learning tends to be highly situation-specific and they rarely transfer spontaneously what they have learned in one place to any new context (Panda, 2001). Research supports the conclusion that these students do not generalise and transfer knowledge and skills unless they are taught in natural environments (Collins, 2007). The implication is that all teachers need to 'teach for transfer' by re-teaching a given skill or concept in different situations (e.g., classroom, workshop, in the community), by explicitly teaching students to recognise where and when a particular skill or strategy can be applied and by positively reinforcing students on any occasion when they show signs of using prior knowledge and generalising previous learning.

Motivation

In the early years, students with mild or moderate intellectual disability usually have normal motivation to learn. However, once in school they tend to experience many difficulties and failures, and the cumulative effect quickly erodes their confidence and they become less and less inclined to put in the required effort (Hallahan et al., 2009). They may become passive and dependent, expecting to be helped by someone (teacher, aide, or peers) or they may become naughty in class to gain a feeling of power that they otherwise lack. Some may develop 'learned helplessness', expecting that anything a teacher asks them to do will be too difficult and will lead to failure.

One of the greatest challenges for teachers and parents of children with intellectual disability is to maintain their motivation and desire to learn. This is achieved most effectively by gearing the tasks we set them and the amount of help we give them to their current abilities, and by using teaching methods that are clear and explicit. The work we set needs to be interesting, appropriate to the student's age, and to present a certain degree of challenge while ensuring a high probability of success. Students with intellectual disability also need frequent praise and reinforcement (if deserved) to increase their self-esteem and feelings of self-efficacy. To help overcome their tendency to blame their difficulties on a personal lack of ability they also need frequent help to reflect upon the positive relationship between the care and effort they expend on tasks and the outcomes they achieve. This process is termed 'attribution retraining'.

Self-management

Self-management (or self-regulation) is important for independent functioning in life. The term refers not only to controlling one's own behaviour and impulses but also to the way one executes daily tasks efficiently by setting relevant goals, managing materials, meeting required expectations and deadlines and using time efficiently. Unfortunately, students with intellectual disability do not develop self-management skills without a great deal of guidance from teachers, parents and peers (Cohen & Spenciner, 2005; Vieillevoye & Grosbois, 2008). Children who have been over-protected in the family environment are particularly weak in this aspect of personal development.

There is clear evidence that these students can be taught self-management to a degree compatible with their level of disability (Westwood, 2007). When students with mild or moderate impairment are taught more effective strategies for self-management, the ease with which they can be successfully included in mainstream classes is greatly enhanced. Of course, students with severe and complex disabilities may never achieve total self-management, but most can be taught to be more self-reliant. In accomplishing this, they need to be given many opportunities to make choices and decisions for themselves, rather than having every minute of their day determined by others (Holverstott, 2005; Neely-Barnes et al., 2008).

Improving social development

Social development is one area where children with intellectual disability may need specific intervention and support. It is a fairly common observation that these students are not always readily accepted into their peer group (and some are openly rejected) largely because they lack the basic social skills that would enable them to relate more easily to other students and to make friends. The situation is most problematic for those students who have speech and language problems and for those with behaviour disorders.

Social skills are the specific behaviours an individual uses to maintain effective interpersonal communication and interaction with others. Social competence therefore comprises a set of skills and behaviours that allow an individual to initiate and maintain positive interactions and cope effectively within the social environment. It is argued that many children with intellectual disability need social skills training and that it should begin in the preschool years when they are most receptive (Cartledge, 2005; Siperstein & Rickards, 2004). Early training in social skills can be instrumental in reducing or preventing problems in later years.

Some research shows that there is fairly limited transfer or maintenance of social skills when they are taught only in contrived training exercises or role play unrelated to the dynamics of the real classroom. Cartledge (2005) advises that social skill instruction should be embedded mainly within the context of events that occur naturally within the child's own classroom or preschool setting. In reality though, to have sufficient impact with older students, training usually needs to combine some individual out-of-class coaching and practice in targeted skills as well as reinforcement of such skills in the real social setting.

Studies suggest that social skills training programs for students with disabilities can be effective if: (i) they target the precise skills and knowledge the individual lacks; (ii) they are intensive and long-term in nature; (iii) they promote maintenance, generalisation and transfer of new skills outside the training context into the individual's daily life (Gresham, 2002; Kavale & Mostert, 2004). The skills targeted for training need to be of immediate functional value to the child in the social environment in which he or she operates.

As well as having appropriate positive pro-social skills, a socially competent individual must also avoid exhibiting negative behavioural characteristics that prevent easy acceptance by others – for example, high levels of irritating behaviour, impulsive and unpredictable reactions, temper tantrums, abusive language or cheating at games. In many cases these undesirable behaviours may need to be reduced by behaviour modification (see Chapter 7).

General principles for teaching intellectually disabled students

So far, several brief references have been made to ways in which teaching strategies may be used to address particular learning characteristics of students with intellectual disability. In this final section, these points are consolidated and some general issues concerning teaching are summarised. In general, the approach needed for all students with intellectual disability is one that carefully combines experiential learning with a necessary amount of direct teaching.

Heward (2009) observes that effective teaching for students with intellectual disability tends to exhibit the following features:

▷ the application of task-analysis that breaks down skills to be learned into achievable steps
▷ a style of instruction that obtains many responses from students in the available time
▷ provision of many prompts, cues, directions and suggestions (termed 'mediated scaffolding') to help a student achieve a learning goal
▷ consistent and frequent use of reinforcement, error correction and feedback
▷ teaching for transfer and generalisation
▷ regular use of formative assessment to check students' progress and to guide instructional design and modifications.

To the above list, Ormrod (2008) would add that new tasks need to be very clearly and completely explained, and all instruction should be carefully paced to ensure a high success rate. Polloway et al. (2008) recommend an appropriate balance between teacher-directed, student-directed and peer-assisted learning.

Reality-based learning

Students with mild or moderate intellectual disability are usually functioning at a concrete operational level of cognition, so they learn best when they are actively involved in the learning process and provided with real situations and real materials (Cohen & Spenciner, 2005; Downing & Eichinger, 2008. This approach is often referred to as 'experiential learning' or 'a hands-on approach'. The more concrete a learning situation is, the more likely it is that students will learn and remember (Drew & Hardman, 2007). For example, rather than teaching concepts of number, money, budgeting and giving change through routine paper-and-pencil exercises and worksheets in class, students are taken instead to the supermarket regularly to buy goods which are then utilised back in the classroom. Similarly, skills in food preparation are learned firsthand by having the students cook meals on certain days in the week and at school camps. Skills in using public transport and accessing community services are also taught in the real world. Reality-based learning of this type results in better generalisation and transfer of knowledge and skills than is achieved with a purely 'chalk-and-talk' instructional approach.

It must be noted, however, that with the most severely handicapped students, particularly those who are highly distractible, it is often necessary to teach new skills first in intensive one-to-one sessions in a controlled and distraction-free setting, before gradually extending the student's program into more natural environments for reality-based experiential learning (Lovaas & Smith, 2003).

Carefully graded and sequenced learning tasks

Vaughn et al. (2007) remind us that students with intellectual disability learn most things at a slower rate than other students, and are particularly confused and frustrated by complex and multi-tiered tasks. As far as possible, learning activities (and indeed the curriculum itself) need to be broken down into manageable steps, taught directly and explicitly, and with much practice and over-learning at each stage (Panda, 2001; Raymond, 2004; Snell & Brown, 2006). Verbal instructions should be accompanied whenever possible by clear demonstrations and visual prompts such as cue cards.

Direct teaching

Direct teaching, based upon behavioural principles of goal setting, task analysis, modelling, shaping through reinforcement, repetition and over-learning has proved to be the most effective method for teaching information and skills. Its value with intellectually disabled students is very well supported by research (Mitchell, 2008; Pressley & McCormick, 1995; Raymond, 2004).

There is evidence that systematic instruction that embodies modelling, prompting and fading techniques works well with these students (Browder et al., 2008). In particular, the *constant time delay method* seems to be valuable for eliciting correct responses from a student. Under the time delay method, a student is presented with a stimulus (for example, a numeral or a word on a card) and asked for a response (e.g., saying the number or word aloud). The teacher waits long enough for the student to think and respond before providing a prompt or correction. This technique can even be taught to peers so that they can employ it in peer tutoring situations with intellectually disabled classmates (Jameson et al., 2008).

Strategy training

Cognitive strategy training (that is, teaching students thoughtful task-approach strategies to use when attempting tasks such as reading comprehension, simple mathematical problem solving, or story writing) can be effective with mildly disabled students (Mitchell, 2008; O'Brien, 2005). However, strategy training is problematic with the moderately and severely disabled because it requires a degree of self-regulation and metacognition that they generally do not possess.

Approaches in early childhood

Allen and Cowdrey (2009) advocate a careful combination of developmentally appropriate activities together with a behavioural approach in the early years. They acknowledge that children with developmental and intellectual delay benefit from a step-by-step curriculum involving both direct instruction and learning through activity and imitation. In the early years, the focus may be mainly on fostering self-care and independence, social skills, language and on managing any challenging behaviours that would reduce the opportunity to learn. However, there are benefits for many disabled children in making an early start on simple academic skills (simple reading and number) rather than adopting a play and activity approach in which such skills are totally ignored (Schickedanz, 1999). There is a need for early childhood teachers to consider developing disabled children's 'readiness behaviours' for learning, such as sitting still, attending to the teacher, following directions, use of pencil and scissors and reacting appropriately to classroom routines (Hallahan et al., 2009).

Issues in adolescence

Teaching protective behaviours

Senior students (male and female) with intellectual disability need to be taught protective behaviour strategies to reduce the possibility that they become the victims of sexual abuse or other forms of exploitation. Their lack of social judgement often causes them to be too trusting and naive. They may not fully comprehend right from wrong in matters of physical contact and are therefore at risk. These students need to be taught that some forms of touching are not acceptable, and they need to appreciate the dangers of going anywhere with strangers or accepting gifts for favours. These students also need to feel there is someone they can talk to confidentially at school if they are ever in such a situation.

The Protective Behaviours Program is operating in all Australian states and in several countries overseas. The approach is based on two principles: first, we all have the right to feel safe at all times. Second, nothing is so awful that we can't talk to someone about it. The approach applies to children of all ages and covers a number of areas other than sexual abuse (for example, bullying, drugs).

Preparing for work

Several observers have remarked that too little attention is given in some schools to preparing senior disabled students for employment (Cronin & Patton, 1993; Fives, 2008; Ormrod, 2008). Since 'holding down a job' is one of the most important achievements that intellectually disabled students must aim for, their schools must address this need by providing appropriate pre-vocational training and pre-work experience (Smith et al.,

2008). This can present a major problem (as yet unresolved) in mainstream secondary schools that offer inclusive education – because a typical high school is not geared to pre-vocational training for students with disabilities. In the past, most senior special schools provided excellent preparation for employment for these students. Regrettably, the move toward inclusion has considerably weakened such provision for older students.

The Links box below provides additional information on topics covered in this chapter. The following chapter looks in a little more detail at some of the specific syndromes associated with intellectual disability, together with other disorders that affect development and learning.

LINKS TO MORE ON INTELLECTUAL DISABILITY

▶ A clear and concise overview of intellectual disability can be found in the Medical Encyclopedia online at: http://www.answers.com/topic/mental-retardation

▶ For more information on students with severe and multiple disabilities, see the website of the National Dissemination Center for Children with Disabilities (US) at: http://old.nichcy.org/pubs/factshe/fs10txt.htm#edimps

▶ Teaching hints for working with intellectually disabled students can be located at: http://www.as.wvu.edu/~scidis/intel.html

Other resources

▶ Copeland, S. R. & Keefe, E. B. (2007). *Effective literacy instruction for students with moderate or severe disabilities*. Baltimore, MD: Brookes.

▶ Drew, C. J., & Hardman, M. L. (2007). *Intellectual disabilities across the lifespan* (9th ed.). Upper Saddle River, NJ: Pearson-Merrill-Prentice Hall.

▶ Polloway, E. A., Patton, J. R., & Serna, L. (2008). *Strategies for teaching learners with special needs* (9th ed.). Upper Saddle River, NJ: Pearson-Merrill-Prentice Hall.

▶ Westling, D. L., & Fox, L. (2009). *Teaching students with severe disabilities*. Upper Saddle River, NJ: Pearson-Merrill-Prentice Hall.

LINKS TO MORE ON PROTECTIVE BEHAVIOURS

▶ For more information on protective behaviours see:
http://www.protectivebehaviours.co.uk/AboutPBsWhatsThat.htm

▶ Useful information on the Protective Behaviours program in Australia at:
http://www.protective-behaviours.org.au/programs.htm

▶ Useful information on the UK program available at:
http://www.protectivebehaviours.co.uk/AboutPBs.htm

Other resources

▶ Harper, G., Hopkinson, P., & McAfee, J. G. (2002). Protective
behaviours: A useful approach to working with people with learning
disabilities. *British Journal of Learning Disabilities*, *30*, 4, 149–152.

▶ Margetts, D. (2002). *Protective behaviours: A personal safety program*.
Canberra: Protective Behaviours Australia Inc.

▶ Rose, J. (2004). Protective behaviours: Safety, confidence and self-
esteem. *Journal of Mental Health Promotion*, *3*, 1, 25–29.

Specific syndromes

▶ Over the past century, several distinct syndromes and disorders that cause learning and developmental problems have been identified.

▶ Teachers' understanding of the nature of these disorders will help them recognise more precisely the specific difficulties and needs of their students.

▶ The diverse nature of the disorders suggests that often a multidisciplinary approach is required, drawing upon the expertise of a range of other professionals in addition to the teacher.

▶ Teaching methods need to take into account the learning and behavioural characteristics of these students.

In this chapter, factual information is provided on some specific syndromes that severely affect learning and development. Some of these syndromes were referred to very briefly in Chapter 1 because they are associated with intellectual disability.

Down syndrome

Down syndrome accounts for the single largest subgroup of individuals with moderate to severe intellectual disability in the school system (Wishart, 2005). Approximately 1 child in every 700 to 1000 live births has the syndrome and it is found in all ethnic groups and in families across the full range of socio-economic and educational status. The risk of having a child with Down syndrome increases with the age of the mother, with a 1 in 2000 chance at age 20, 1 in 1000 at age 30, but a 1 in 20 chance at ages above 45 (Roizen, 2007).

This form of impairment has been known since ancient times, but was eventually identified as a specific syndrome in the mid 1800s. The underlying genetic cause of the condition, a chromosomal abnormality, was not confirmed until 1959. In the vast majority of cases, Down syndrome is due to an extra chromosome 21 (three instead of the normal two). The condition is also referred to as Trisomy 21.

Children with Down syndrome are readily recognised by their physical features, particularly their facial characteristics with upward slanting eyes and snub nose. Pictures of smiling, happy children with Down syndrome are often used in charity advertising and for fundraising purposes because most people in the community readily associate their physical appearance with intellectual disability.

Despite the impression that all Down syndrome children are happy, friendly and affectionate, they are actually as varied as any other group of individuals in terms of personality, abilities, interests, behaviour, social skills and self-management (Wishart, 2005). The social responsiveness often observed in Down syndrome children, particularly if they have been raised in a loving and supportive family, can lead to an incorrect assumption that their cognitive ability is higher than it actually is. Within the group there is a wide range of IQs. Some children are regarded as 'high functioning' and are within the mildly disabled range, but many others (85 per cent) are moderately to severely impaired (Roizen, 2007).

More than 44 per cent of children with Down syndrome also have heart conditions, 60 per cent have vision problems and 66 per cent have hearing defects. Some suffer from sleep apnoea as a result of nocturnal breathing problems, and through being overweight. In general, these children seem to have a lowered resistance to infection. Life expectancy for individuals with Down syndrome was only about 20 years in the 1930s, but with improved health care and other support it has now increased to 50 years.

In terms of development, not many Down syndrome children with moderate cognitive impairment advance beyond a mental age of 8 years, but through high-quality teaching and training most can acquire some degree of independent functioning (Wishart, 2005). It is said that children who are raised in a supportive home environment make much better progress than those raised in institutions. Higher-functioning adults with the syndrome can sometimes hold down a simple job if they are thoroughly trained and given ongoing supervision. Many, however, find themselves in sheltered workshops, activity centres or similar situations, or may have to remain at home or in an institution. Several experts have commented on the fact that there is a tendency for individuals with Down syndrome to lose some of their skills and cognitive abilities as they move from adolescence into adulthood, although they may still continue to acquire additional skills in the areas of social competence and everyday living (Hallahan et al., 2009; Hodapp & Dykens, 2003; Roizen, 2007).

Higher-functioning children with Down syndrome are now frequently placed in regular preschools and primary schools where they are provided with additional support and usually have an individual education plan (IEP) prepared for them. Wishart (2005, p. 82) writes: 'Some children [with Down syndrome] do comparatively well in school, but average academic achievement levels remain low'. Inclusion of Down syndrome students at upper primary and secondary level becomes far more problematic because of their weaker mental ability and greatly reduced language competence. Their lack of foresight and self-management also pose a major challenge (Lovering & Percy, 2007).

Parents are routinely directed toward early intervention programs (if they are available in their area) soon after the child is born. Evaluation of early intervention programs suggests that they do have some value in the short term in helping with motor coordination, speech and socialisation, but response varies significantly from child to child (Hines & Bennett, 1996). Wishart (2005) comments that investment in early intervention and special education has so far failed to produce substantive or lasting benefits for the majority of Down syndrome children.

In terms of learning characteristics, children with Down syndrome (like most other intellectually disabled students) are reported to be relatively better at learning through visual and concrete methods rather than through auditory means (Frenkel & Bourdin, 2009). When teaching these children, verbal instruction needs to be supported by visual cues (symbols, pictures or objects) and very clear demonstrations need to be given when teaching particular skills. It is also helpful to simplify and shorten verbal instructions as much as possible and to ask the child to repeat important instructions back to you. Some of these children are said to have one or two areas of relative strength (e.g., word recognition) (Cardoso-Martins et al., 2009; Kliewer, 1998; Roch & Levorato, 2009). These strengths are particularly found in the case of some higher-functioning individuals and can be used by the teacher to boost the child's self-esteem and motivation.

Learning for many Down syndrome children is made difficult by their poor span of attention and their tendency to be distracted easily. Some have attention deficit hyperactivity disorder (ADHD) which compounds their learning problems (Roizen, 2007). Poor working memory is often cited as one of their major weaknesses, causing them significant difficulties when processing sequences of information, solving problems or dealing with two or more sources of information at once.

Fragile X syndrome

Fragile X syndrome was formally identified in 1991, although studies had begun to yield data on the disorder as early as the 1940s. The prevalence of this genetic disorder is about 1 in 2000 live births (Hagerman, 2004), affecting males more frequently and more severely than females. Approximately 15 per cent of males with fragile X syndrome are considered high-functioning and may have intelligence within the normal range. However, the majority have mild to moderate learning difficulties, with a few exhibiting a severe degree of disability. It is now believed that an estimated 5 to 7 per cent of intellectual disability in males may be due to the fragile X factor (Allen & Schwartz, 2000). Females with fragile X are more likely to have only mild intellectual disability or general learning difficulties; but some 30 per cent of affected females show signs of emotional disturbance and abnormal speech patterns.

Common problems in this disorder include language delay, behaviour problems and autistic-like responses (poor eye contact, hand flapping, hyperactivity, poor sensory skills). While most children with fragile X enjoy social interaction, about 1 in 3 shows some signs of autism (e.g., social withdrawal, perseveration in speech, thought and behaviour, stereotypic habits). Braden (2004) reports that children with fragile X tend to be inflexible in their thinking and over-selective in the stimuli they choose to attend to. They seem to have a desire for sameness and routine (again, suggesting autism) and they often display obsessive behaviour such as a fixation to complete any task they are given before moving on to something new. If forced to stop to attend to something new, they may become anxious or adopt avoidance or non-compliant behaviour. Anxiety is seen as a common trait in children with fragile X syndrome (Woodcock et al., 2009).

According to Cornish (2004) the main learning problems evident in many of these children relate to inattention, impulsiveness and hyper-activity. This may account for the fact that boys with fragile X tend to have difficulty when confronted with tasks that require sustained concentration and careful step-by-step processing (e.g., mathematical calculations) (Hodapp & Dykens, 2003).

Students with this disorder benefit most from a highly structured learning environment and systematic instruction. Computer-assisted learning can be effective with some students by holding their attention and providing visual input together with active responding. Some students will require carefully planned behaviour management programs. Dew-Hughes (2004) observes that educating a child with fragile X syndrome challenges the professional knowledge and practice of teachers who must respond to each child's unique needs through differentiated instruction.

Prader-Willi syndrome

The prevalence of this relatively rare chromosomal disorder (usually invol-ving chromosome 15) is approximately 1 in 15 000 live births (Hodapp & Dykens, 2003), although some experts report it be as low as 1 in 25 000 (Gordon, 2007). It was first identified in 1956.

Children with Prader-Willi syndrome exhibit two quite different phases during their development. Infants go through a period in which they fail to thrive due mainly to poor muscle tone and feeding and sucking difficulties. It is recommended that these children should receive physiotherapy to help improve muscle tone. For some, speech therapy may also be required. Later, but before the age of 6 years, these children begin to develop more rapidly. However, at the same time there is a marked tendency for them to acquire a number of maladaptive behaviours including temper tantrums, stubbornness, obsessive-compulsive habits, aggression and stealing (Kim et al., 2005; Reddy & Pfeiffer, 2007; Woodcock et al., 2009). Their most serious problem is the development of a voracious appetite (*hyperphagia*) leading easily to obesity. They will often expend much effort searching for and stealing food. Research seems to indicate that this insatiable overeating may result from a lack of feeling of 'fullness' due to some abnormality in the hypothalamic region of the brain. In other words, the eating disorder has organic causes rather than being a maladaptive learned behaviour. One of the main management priorities in home and school is to control the child's food intake.

Work by Curfs and Frym (1992) suggests that some 32 per cent of persons with Prader-Willi syndrome have IQs above 70 (i.e., they are not intellectually disabled). Further research has found that 34 per cent of Prader-Willi sufferers could be classified as being in the mild range of intellectual disability, 32 per cent in the moderate to severe range, and about 1 per cent profoundly disabled (Gordon, 2007).

For most children with this syndrome, learning difficulties occur most obviously in areas of auditory information processing, short-term memory, attention, arithmetic, and writing. Inappropriate behaviour can also disrupt learning and can lead to socialisation problems. As with most other forms of intellectual disability, students with this disorder need a highly controlled and predictable learning environment.

Foetal alcohol spectrum disorders (FASD)

Excessive alcohol consumption by the mother during pregnancy is now known to cause a wide range of adverse physical and neurological conditions in the unborn child (Davidson & Meyers, 2007; O'Leary, 2004). These disorders produce what has been termed foetal alcohol syndrome (FAS) or, in a mild form, foetal alcohol effects (FAE). Mothers who have had one child with FAS stand almost a 50 per cent chance of having another child with the same disorder unless they significantly reduce their drinking habit (Batshaw et al., 2007). Weiten (2001) suggests that even a social level of drinking can have enduring effects on the development of the foetus. The prevalence of FAS is estimated to be approximately 1 per 1000 live births, with another 3 to 5 per 1000 showing some effects (FAE).

The most marked features of FAS include low birth weight, a failure to thrive in the first years of life, retarded physical development, abnormalities of the central nervous system, delayed speech development and learning difficulties ranging from mild to severe (Davidson & Meyers, 2007). It is also observed that many (but by no means all) of these children have a smaller head size than normal, and share a characteristic facial appearance, with widely spaced eyes, narrow eyelids, an upturned nose and thin upper lip. Up to 50 per cent have additional handicaps in vision and hearing, or physical disorders including heart problems and abnormalities in limbs and joints (Weiten, 2001).

The home environment of many children with FAS and FAE is often very far from ideal, particularly if the mother is a chronic alcoholic (as is most often the case). Emotional, behavioural and learning problems can be greatly exacerbated by the context in which the child is reared. Behaviour problems are quite common in these children, and can be a major source of concern in both the school and the family context. Hyperactivity and impulsivity are observed in many cases, creating further barriers to learning in school (Nicholson, 2008).

The intelligence level of children with FAS and FAE varies greatly, ranging from normal intelligence to moderate or severe cognitive disability. The majority of cases tend to fall within the 'slow learner' and mild disability categories. These children are usually accommodated within the mainstream education system, but often require intensive remedial teaching in basic skills (Dybdahl & Ryan, 2009). For children with FAS and intellectual disability placed in special schools, teaching methods correspond with those described in Chapter 1.

Turner syndrome

This disorder affects only females. The prevalence rate is reported to be 1 in 2500. Those affected have short stature, with a broad and abnormally short neck and wide chest. Congenital heart conditions are common, together with frequent ear infections and hearing loss. Puberty is usually delayed in these girls, and they remain infertile. General intelligence may be average, but many learning difficulties are evident, usually due to visual perceptual problems (Gordon, 2007).

Girls with Turner syndrome often require remedial teaching and, if hearing loss is significant, may also benefit from teaching adaptations described later in Chapter 5.

Williams syndrome

Williams syndrome is a rare disorder in which the children affected tend to have a characteristic facial appearance (often described as 'elfin') and are often very talkative and outgoing. Their oral language skills tend to be very much above their other abilities (Hodapp & Dykens, 2003). In most cases, the degree of intellectual disability falls within the mild to moderate range, and most of these children have additional health problems (including cardiovascular) that require close monitoring. The prevalence rate is reported to be 1 in 20 000 (Gordon, 2007; Semel & Rosner, 2003).

Teaching methods for these students depend on the degree of cognitive impairment. Those with intellectual disability require a carefully structured curriculum and direct instruction. Remedial teaching is also necessary for those who are frequently absent from school for health reasons.

LINKS TO MORE ON SPECIFIC SYNDROMES AND DISORDERS

Down syndrome
▶ http://kidshealth.org/parent/medical/genetic/down_syndrome.html

Fragile X syndrome
▶ http://www.medicinenet.com/fragile_x_syndrome/article.htm

Prader-Willi syndrome
▶ http://www.nlm.nih.gov/medlineplus/praderwillisyndrome.html

Foetal alcohol syndrome
▶ The Academy of American Family Physicians website carries a comprehensive article at: http://www.aafp.org/afp/20050715/279.html

Turner syndrome
▶ http://kidshealth.org/teen/diseases_conditions/genetic/turner.html

Williams syndrome
▶ http://www.ninds.nih.gov/disorders/williams/williams.htm

Pervasive developmental disorders

▶ Pervasive developmental disorders represent a low-incidence category of disability, but children with these disorders present an enormous challenge to their parents, teachers and other professionals.

▶ Autism is the most widely recognised pervasive developmental disorder, characterised by significant deficits in communication, impaired learning capacity, problems with socialisation and unusual patterns of behaviour.

▶ Many treatments and teaching approaches have been developed for autistic children. Some focus on improving social skills and communication, others on modifying behaviour and increasing parental involvement.

▶ Students with Asperger syndrome are higher functioning than children with classic autism. They can commonly be educated in ordinary classes with support.

▶ Rett syndrome and childhood disintegrative disorder are two fairly rare conditions that share some characteristics of autism.

Pervasive developmental disorders (PDDs) – sometimes referred to as disorders of social communication – include autism, Asperger syndrome, Rett syndrome and childhood disintegrative disorder. Also included are individuals who exhibit some abnormal development in social and communicative skills but whose problems are not so severe or so clearly delineated that they can be placed with certainty in any of the above categories. This sub-group is given the cumbersome classification 'pervasive developmental disorder not otherwise specified' (PDD-NOS).

Pervasive developmental disorders vary in severity and are generally regarded as neurodevelopmental in nature (i.e., associated with the developing brain) (Carper, 2004). The underlying cause is not yet known, but research is ongoing in this area. Although the overall prevalence rate for all forms of PDDs combined is reported to be approximately 3 per 1000 births, these disorders account for some 25 per cent of children with moderate to severe intellectual disability (Chiu et al., 2008). These children present an enormous challenge to their parents, teachers and caregivers because of their low level of adaptive behaviour and their general lack of self-management. They are said to have a 'triad of impairments' comprising social deficits, language and communication deficits, and atypical behaviours (Neisworth & Wolfe, 2005). Some children with PDDs have no speech at all, or may use repetitive language patterns, repeating what others say but without understanding (echolalia). Similarly, they have a limited understanding of instructions. It is also observed that some individuals with severe PDDs tend to regress in their language skills over time (Meilleur & Fombonne, 2009). It is suggested that these language problems may be due in part to an underlying auditory processing disorder (see Chapter 6).

Many children with PDDs also have emotional and behavioural problems, including hyperactivity, temper tantrums, aggression or extreme withdrawal, phobias, and a tendency toward self-injury (Cooper et al., 2009). However, children with Asperger syndrome and those with PDD-NOS tend to be higher functioning, more responsive to intervention, and achieve a better level of adaptive behaviour and independence.

Autism

Symptoms of autism are typically identified in infancy (prior to 3 years of age). These symptoms vary in frequency and intensity. Typical symptoms include:

▶ difficulties understanding and using language
▶ greatly reduced ability to learn, particularly through observation and imitation

- lack of interest in relating to other people or forming emotional bonds
- avoidance of eye contact
- unusual patterns of play (or the child may not play at all, instead engaging in repetitive and stereotypic movements and rituals such as hand-flapping or rocking)
- obsessive interests
- lack of imagination and initiative
- resistance to change in daily routines
- hypersensitivity to environmental stimuli (e.g., loud noises, or flickering fluorescent lights) (Hegde, 2008).

Individuals with autism have this disorder to varying degrees and any single child diagnosed as autistic may not show all the above characteristics. To be officially diagnosed as autistic, a child must show symptoms of abnormal linguistic, cognitive and social development before the age of 3 years and must meet at least 6 of 12 criteria listed in the *Diagnostic and Statistical Manual of Mental Disorders* (APA, 2000). The manual delineates in detail the key areas of abnormal development believed to be typical of children with autistic spectrum disorders.

Some students with mild autism are close to normal in many facets of their behaviour, but others with more severe autism are very low-functioning in terms of cognition, self-regulation and social development. Some children with moderate to severe degrees of autism often sit for hours engaging in unusual repetitive behaviours (*stereotypic behaviours*) such as spinning or flicking objects, flapping their fingers in front of their eyes, or just staring at their hands (Woodyatt & Sigafoos, 1999). Some autistic children exhibit self-injurious behaviour such as biting or scratching their arms or hands. It is believed that these stereotypic behaviours may be triggered by anxiety or by a need for additional sensory stimulation (Joosten et al., 2009). It is generally agreed that children with autism spectrum disorders tend to exhibit higher than normal levels of anxiety, particularly when faced with changes to normal routines (MacNeil et al., 2009).

Autism is a low-incidence disability with a ratio of males to females of 5 to 1. There is a suggestion that the prevalence of autism is increasing, but this is probably a false conclusion due to the fact that there is an increasing awareness of the disorder and to the use of a broader definition of various forms of disability that fall within the autistic spectrum (Holborn, 2008).

Autistic children have been identified in all parts of the world and the disorder does not appear to be culturally determined. Although autism has been found in individuals at all levels of intelligence, 75 per cent of children with autistic disorders have IQ scores below 70 and require ongoing intensive special education and behaviour management. Children with autism remain among the most difficult students to place in mainstream classrooms (Humphrey, 2008; Turnbull et al., 2007). In the US, Australasia and Britain, only about 12 per cent of children with diagnosed autism currently receive their education in mainstream settings.

Treatment, training and education in autism

Although there is no cure for autism, or for any other PDD, individuals can be helped to perform at a higher level if provided with high quality intensive teaching and management (Hyman & Towbin, 2007; Wong & Westwood, 2002). However, progress is usually very slow. Nevertheless, the benefits reported for many of these children are positive enough to make the investment of time and effort worthwhile (Cipani, 2008). Children with mild forms of autism can often be accommodated in mainstream classrooms with support; but children with more severe forms usually have to be placed in special schools, where their needs can be more effectively met with an individualised program and by well-trained teachers.

Individualised teaching sessions for children with autism need to be implemented very frequently and should follow a predictable schedule. Studies have strongly supported a view that the most effective intervention strategies are highly structured and delivered with intensity (Waterhouse, 2000). New information, skills or behaviours must be taught in small steps using consistent, systematic and direct methods. Each child's program must be based on a detailed appraisal of his or her current developmental level and existing repertoire of skills and responses. Program goals need to build upon relative strengths. All teachers, parents and other caregivers should be familiar with the precise goals and objectives of the program and must collaborate closely on methods to be used with the child.

It is essential that parents be very familiar with the teaching strategies used in any intervention program so that teaching can be continued consistently in the home environment. The most effective interventions involve the child's family as well as teachers and therapists. Home-based intervention programs (or programs combining school- or clinic-based intervention with home programs) produce the most enduring results.

Many of the general principles for teaching students with intellectual disability, as summarised in Chapter 1, apply equally to children with autism. In addition, several specific approaches have been developed, including some that remain controversial and of doubtful value (Levy & Hyman, 2005). Effective intervention approaches include programs with the following emphases (Prelock, 2006):

▶ *Relationship-based intervention.* This approach is designed to facilitate and increase the autistic child's social and emotional attachment to others (parents, caregivers, teachers, paraprofessionals and peers). Examples include play therapy, holding therapy, DIR/floortime approach, social skills training, social stories and Relationship Development Intervention. These approaches are described later.

▶ *Skills-based intervention.* This approach is designed to teach important self-help and self-management skills and strategies, and to develop some basic language, literacy and numeracy skills. Verbal Behaviour Intervention approach falls into this category. Usually direct instruction methods are employed for this purpose, but every opportunity is also taken to teach and reinforce skills when 'teachable moments' occur naturally during the day. Discrete Trial Training (DTT) is frequently used in which a specific response is targeted, directly taught, practised intensively, reinforced and eventually produced by the child on signal (de Boer, 2007) (see Links box for details). This technique is much used in the Lovaas program for autistic children, as described below (e.g., Lovaas, 1993; Lovaas & Smith, 2003).

▶ *Physiologically-oriented intervention.* This may include the use of medication, controlled diet, sensory training, and relaxation therapy (e.g., music therapy, Snoezelen). For detailed information on diet and medication in autism see the Links box at the end of the chapter.

▶ *Multidimensional approach.* This combines several of the above emphases and also includes family counselling, parent training and involvement and community support. Examples include TEACCH and SCERTS (see below).

Most intervention approaches aim to increase the child's ability to communicate, improve his or her motivation to interact socially with others, teach some functional skills, and reduce negative behaviours by replacing them with more useful responses. Stereotypic behaviours need to be reduced as far as possible, in order to make the child more available for new learning and to increase the chances of social acceptance by peers (Conroy et al., 2005). Behaviour modification techniques using principles of applied behaviour analysis (goal setting, direct instruction, modelling, reinforcement, and time out) have proved to be the most effective methods for changing behaviour patterns and developing some degree of self-management (Dempsey & Foreman, 2001; Hyman & Towbin, 2007; Scheuermann & Webber, 2002).

Cognitive training approaches that set out to teach autistic children self-regulating strategies (e.g., using pre-rehearsed scripts of 'self-talk' to control one's own behaviour) can be effective with high-functioning autistic children and those with Asperger syndrome. However, this approach is problematic with low-functioning children because it requires self-monitoring and metacognition not usually found in these children.

One of the main priorities for autistic children is to teach them to play – something that comes naturally to other children (Beyer & Gammeltoft, 2000). Play is essential for physical, cognitive, social and emotional development. More will be said later on this issue under the sections covering play and play therapy.

Using alternative methods of communication

Communicating effectively with autistic children is one of the top priorities for teachers and parents. It is commonly agreed that most of these children are able to process visual information much better than verbal information. For non-verbal autistic children, systematic use of visual cues (hand signing, gesturing, pointing, picture cards and symbols) is usually necessary in most teaching situations to accompany verbal instructions and to enable the child to respond (Adams, 2008; Shane & Weiss-Kapp, 2008). For some autistic students, personalised picture boards and other forms of visual communication aid can be of great value. One of the most widely adopted supplementary methods in special schools is the Picture Exchange Communication System (PECS) (Frost & Bondy, 1994). Using these materials a child can make his or her requests known by passing an appropriate card (e.g., a picture of a drink) to the teacher or parent, and can respond to specific questions from the teacher by selecting an appropriate picture. For further details of PECS, see the Links box.

Specific interventions and methods

This section describes a broad range of methods and approaches that are currently being used with autistic children. As indicated by Prelock (2006), some interventions focus on communication and social development, skill building, and behaviour change, while others are of a more therapeutic nature. Much more information on each method or approach can be located online using the links provided at the end of the chapter.

DIR/floortime approach

The letters DIR represent *developmental, individualised, and relationship-based*, these being the three underlying principles of this early-years approach (Greenspan & Wieder, 2006). The term 'floortime' refers to informal sessions literally 'on the mat' that engage the child and an adult in natural interaction and close contact through the medium of developmentally appropriate and enjoyable play activities. The aim of DIR/floortime is to improve the child's overall ability to communicate, think and relate to others. The approach is implemented chiefly by parents at home, but it can be adapted by teachers and paraprofessionals in a preschool context.

Relationship development intervention

This is a family-based treatment that places emphasis on helping the child develop a better understanding of other people and thus acquire empathy, an ability to express affection, and willingness to share experiences and activities with others (Gutstein et al., 2007, Gutstein & Sheely, 2002). The approach assumes that children with autism have the potential to achieve better social adjustment if they are exposed to social situations in a gradual and systematic manner and if they have these situations explicitly interpreted for them. Gutstein theorises that the typical behaviours and difficulties associated with autism are due in part to 'neural underconductivity' and that this can be overcome to some extent by providing appropriate experiences. Parents involved in this approach receive training from professionals to maximise their own ability to communicate effectively with their child. Ongoing professional support is then provided.

TEACCH approach

The acronym TEACCH stands for Treatment and Education of Autistic and Communication-handicapped Children (Mesibov et al., 2005). This multi-tiered approach stresses the need for a high degree of structure in the autistic child's day and environment. It uses a combination of cognitive and behavioural-change strategies, coupled with direct teaching of specific skills. Importance is placed on training parents to work with their own child and to make effective use of available support services. An important feature of the approach is that it capitalises on autistic children's preference for a visual mode of communication (e.g., sign language, picture cues) rather than the auditory-verbal mode. In general, TEACCH has proved to be of positive value in the education of autistic children (Simpson, 2005).

SCERTS model

The acronym SCERTS is derived from Social Communication, Emotional Regulation and Transactional Support; and these are the areas of development prioritised within this transdisciplinary and family-centred approach (Prizant et al., 2005). The overriding goal of SCERTS is to help a child with autism become a more competent participant in social activities by enhancing his or her capacity for attention, reciprocity, expression of emotion and understanding of others' emotions. In particular, SCERTS aims to help children become better communicators and to enhance their abilities for pretend play. SCERTS attempts to capitalise on naturally occurring opportunities for development throughout a child's daily activities and across social contacts including parents, siblings, other children and caregivers. The designers of this model stress that other treatment and teaching methods can be integrated into SCERTS. For more information see the Links box.

Son-Rise

This program encourages parents to adopt a positive attitude toward their autistic child's potential progress, and it provides advice on how to become effective therapists and teachers of their own children (Kaufman, 1995; Williams & Wishart, 2003). Son-Rise uses mainly the home as the basic setting for non-directive intervention. The underlying approach requires parents and other caregivers to 'go with the child' in his or her behaviours rather than opposing and disciplining the child by aversive means. Use the child's existing strengths at all times, even joining with the child in, for example, a repetitive and ritualistic behaviour and gradually attempting to shape it into something more positive. This aspect mirrors some of the principles of 'intensive interaction approach', already discussed in relation to intellectual disability (see Chapter 1).

Lovaas program

One very intensive program for autistic children is that devised by Lovaas (1993; Lovaas & Smith, 2003). The program begins with the child at age 2 years and involves language training, teaching social behaviours and the stimulation of play activity. Emphasis is also given to the elimination of excessive ritualistic behaviour, temper tantrums and aggression. The second year of treatment focuses on higher levels of language stimulation and on cooperative play and interaction with peers. Lovaas claims high success rates for the program, including increases in IQ. He claims that almost half of the treated group of children reached normal functioning levels. While the general principles are undoubtedly sound, the fact that this program takes 40 hours per week over two years using one-to-one teaching makes it very labour-intensive and expensive. It is difficult if not impossible to replicate the whole approach in the average special school or preschool.

Miller Method

This approach encourages the autistic child to make explorations, thus increasing his or her awareness of the environment and assisting communication and concept development (Miller & Chretien, 2007). Inappropriate behaviours are eliminated and replaced with more productive behaviours. Sign language is used to supplement spoken language. An unusual feature of this approach is that, unlike DIR/floortime where the child is usually worked with on the floor or mat, in the Miller Method, the child is often placed in a slightly elevated position on a box or small platform to improve balance, confidence, body image, and to make eye contact easier.

Verbal behaviour intervention

This approach, also known as the 'applied verbal behaviour approach' uses behavioural principles to teach and maintain essential speech and language skills such as naming objects, making requests, asking questions, replying and conversing. The underlying principles of the approach are derived from Skinner's (1957) analysis of verbal behaviour in which he attempts to explain the way human language is acquired and shaped (Barbera & Rasmussen, 2007; Carr & Firth, 2005). Results suggest that the approach is moderately successful; but unless every effort is made to help the child with autism acquire new language in a natural situation rather than a clinical training setting, there is little likelihood of skills transferring and generalising. Objectives are best achieved by using both direct instructional methods and by maximising the naturally occurring opportunities in the child's daily life.

Pivotal response training

This approach has been developed by Robert and Lynn Koegel (2006) and is based on the principle that intervention should focus on strengthening particular behaviours that will simultaneously have a beneficial effect on other associated behaviours. These 'pivotal behaviours' in a child's repertoire are those that have widespread positive effects leading to generalised improvement. An example of a pivotal behaviour is the autistic child's often erratic responses to multiple stimuli. These inappropriate responses impair the child's ability to learn and to attend to what is relevant. Pivotal response training seeks to reduce inappropriate responses, increase the child's selective attention and improve the ability to combine information from different sources, thus enhancing learning. Pivotal response training employs basic applied behaviour analysis techniques and has been used effectively in the areas of language skills, play and social behaviours.

Motivation is another pivotal response, because improved motivation increases attention to task, participation and autonomy. Training to improve motivation might include the use of preferred activities and opportunities for the child to make many choices.

Play and play therapy

Play is an essential and normal part of childhood, contributing enormously to cognitive, linguistic and social development. For most children, play activities are self-initiated and spontaneous but many children with autism have major deficits in play behaviour (Woodward & Hogenboom, 2000). A priority for young autistic children is therefore to teach them to play. Simple games that involve taking a turn, planning ahead, making a request, taking action (such as picking up or putting down a card or moving a counter) are useful staring points and can stimulate communication, both verbal and nonverbal (Macintyre, 2002). Activities involving visual materials (toys, picture cards, objects) are particularly recommended. Puppet games can be used in association with a 'social stories' approach (see below) to engage and maintain children's attention. Some forms of play therapy used with children who are not autistic claim to be 'nondirective', following the child's lead rather than having the teacher or therapist impose structure on the situation with specific goals in mind. However, in working with an autistic child, it is usually necessary to operate in a much more structured manner by having specific learning goals in mind, providing modelling, imitation of responses and frequent reinforcement.

Music therapy

Music therapy is frequently used with autistic children because it can provide enjoyment and relaxation while at the same time presenting many opportunities for participation, movement and communication (Jordan, 2001). In particular, music activities can create opportunities for social interaction with other children. Music and songs can even be incorporated into the routines involved in teaching self-care skills such as washing hands, cleaning teeth, feeding and toileting (Kern et al., 2007). To be effective, music therapy needs to be provided by a qualified and experienced therapist who structures the sessions to develop the child's cognitive, affective and communicative abilities, and to control negative behaviours such as aggression or hyperactivity (Woodward & Hogenboom, 2000).

Sensory integration therapy

This approach aims to help the autistic child process sensory input more efficiently to avoid, for example, extreme reactions to loud sounds or bright lights, and to enable the child to make better sense of the environment. By improving sensory integration, it is hoped that learning will be enhanced through better attention, concentration, listening and self-control. The approach involves a program of sensory stimulation activities and gross and fine motor exercises. The program is usually carried out by an occupational therapist three or four times a week. However, there is no clear evidence that this approach brings about any significant long-term changes, although often the activities are enjoyed by the child (Reynolds, 2000).

Social stories

A technique that appears helpful in developing autistic children's awareness of normal codes of behaviour is the use of 'social stories' (Crozier & Sileo,

2005; Howley & Arnold, 2005; Quirmbach et al., 2009). Social stories are simple narratives, personalised to suit the child's own needs and behaviours, and to which the child can relate. The theme and context of each story helps the autistic child perceive, interpret and respond more appropriately to typical social situations – for example, sharing a toy, taking turns or standing in line (Earles-Vollrath et al., 2006). Social stories are also being used with young children who are not autistic but who have other serious emotional and behavioural problems (Haggerty et al., 2005).

The book *Autism spectrum disorders: Issues in assessment and intervention* by Prelock (2006) contains valuable information on several other approaches. Each approach is clearly described and evaluated for its effectiveness.

Asperger syndrome

This pervasive developmental disorder was first identified in the 1940s by Hans Asperger, an Austrian physician. It affects males much more frequently than females. Students with the disorder have some of the behavioural and social difficulties associated with autism but to a milder degree; and they tend to have language and cognitive skills in the average or even above average range. Most students with Asperger syndrome are not intellectually disabled and are educated in mainstream schools. Their unusual social behaviour patterns and communication style often cause them to be regarded as strange or eccentric by their peers and teachers; and they often have difficulty making and keeping friends. Hegde (2008) suggests that they lack social tact and a sense of social appropriateness. Similarly, Le Roux et al. (1998, p. 123) comment:

> These children lack an understanding of human relationships and the rules of social convention. They are unusually naive and conspicuously lacking in common sense. Their inability to cope with or adjust to social convention often causes them to be easily stressed and emotionally vulnerable.

Some researchers argue that Asperger syndrome simply represents the higher-functioning end of the autistic spectrum. Others believe, however, that Asperger syndrome is a different form of disability representing a discrete group of individuals with only a few autistic tendencies (Baker & Welkowitz, 2005).

Students with Asperger syndrome vary considerably one from another, but they tend to exhibit some of the following characteristics:

▶ *Egocentrism.* They have difficulty appreciating the views and feelings of others (Adams, 2006).

▶ *Atypical language style.* They may over-use stereotypic phrases; they may repeatedly ask the same question of the same person even though it has been answered; they often use a pedantic style of speech.

▶ *Emotional vulnerability.* They can become anxious and distressed even by fairly minor events.

▶ *Attention deficits.* Their concentration span during typical classroom activities tends to be short. They are easily distracted. However, when engaged in their own interests (or obsessions) they are able to concentrate very well for long periods.

▶ *Memory.* Despite their attention problems, these students often display remarkable memory for facts and information on the restricted range of topics in which they are intensely (sometimes obsessively) interested.

▶ *Academic underachievement.* Although many of the students acquire basic literacy and numeracy skills they seem unable to apply higher-order thinking and reasoning in most areas of the curriculum.

▶ *Social difficulties.* They do not relate easily to other students and they are sometimes the target for teasing or bullying.

▶ *Physical awkwardness.* Many of these students appear to be poorly coordinated.

Teaching students with Asperger syndrome

The key educational considerations for these students include:

▶ strategic seating in the classroom so that they can be monitored closely and kept on task

▶ using very clear instructions, and checking for understanding at all times

▶ direct, literal questioning rather than open-ended questioning

▶ trying to establish reasonably predictable routines and structure to all lessons

▶ using visual aids during lessons wherever possible

▶ allowing more time for these students to complete work

▶ if feasible, using a student's obsessive interests as a focus for some schoolwork, while trying to extend and vary the student's range of interests over time

▶ avoiding the over-use of complex language that requires deeper interpretation (e.g., metaphors, idioms). These students tend to use and interpret language in a very literal manner (Humphrey, 2008).

Some students with this form of autism may benefit from personal counselling to discuss issues such as the feelings of others, social interactions, dealing with their own problems and how to avoid trouble with other students and teachers. The student's own inability to understand the emotional world of others will not easily be overcome but he or she can be taught some coping strategies (Le Roux et al., 1998). Adams (2006) suggests that autistic children and those with Asperger syndrome lack a 'theory of mind' – meaning that they do not seem to be aware of other people's thinking, or to understand the mental states and feelings of others. Counselling should aim to help these students develop normal empathic understandings so that they can interpret the behaviour and intentions of others more accurately. Children with Asperger syndrome are stronger candidates than lower-functioning autistic children for social skills training and self-management training (Choi & Nieminen, 2008). There is a real need to help these students participate successfully in cooperative group work and interact positively with their peers (Eggett et al., 2008).

Self-management training can be regarded as a form of cognitive behaviour modification technique where the student is taught to use mental 'self-talk' to help him or her control responses and reactions appropriately (Ganz et al., 2006; Westwood, 2007). The use of cue cards taped to the student's desk can be useful to remind the student to use appropriate behaviour and to give appropriate responses (McConnell & Ryser, 2005). Teaching self-management is a much more powerful method (and the results are transferred far more readily) than having an adult constantly controlling, or attempting to modify, the individual's behaviour.

Rett syndrome

Rett syndrome is classed as a pervasive developmental disorder and shares some behavioural characteristics with autism. The disorder is due to mutations in a specific gene within the X chromosome (Gordon, 2007). Rett syndrome is found almost exclusively among females because males with the disorder tend to die before birth or within the first year of life.

Prevalence of Rett syndrome is less than 1 in 12 000 births (Gordon, 2007). Life expectancy is now 40 years.

Rett syndrome is a progressive disorder with its most destructive period being between the ages of 1 and 4 years. The child may appear normal at birth and may begin to develop normally, but then starts to regress significantly in cognitive abilities, language and motor skills (Hyman & Towbin, 2007). There are a few mild forms of the condition, but most individuals with the syndrome develop autistic behaviours such as stereotypic hand movements, together with spasticity of the lower limbs. In many cases the child loses the ability to walk. Seizures are common, together with intellectual disability. Most of the girls with this disorder have only limited (or no) verbal skills and usually need to be taught by using alternative modes of communication (sign language, picture cues).

Teaching methods and supports are similar to those needed for autistic children with intellectual disability, as described above and in Chapter 1.

Childhood disintegrative disorder

This relatively rare condition is also referred to as Heller syndrome. The child develops normally in the early preschool years, but from around the age of 4 or earlier begins to lose most of the language, communication and social skills previously acquired. Often self-care abilities, including feeding and toileting are also lost. After a time (usually by age 10) the regression stops, but the child does not regain the lost skills. At first it appears as if the child is developing autism (and some children with the condition have been incorrectly diagnosed as autistic); however, this disorder differs from autism in its pattern of onset, course of development and its outcomes (Hyman & Towbin, 2007). The cause of the disorder is as yet unknown.

It is suggested that intervention methods should be the same as those found effective with autistic children, namely skills-based systematic training using mainly behavioural methods and reinforcement, while at the same time attempting to facilitate the child's emotional and social connectedness to others (Adams, 2008; Prelock, 2006; Shane & Weiss-Kapp, 2008). Non-verbal methods of communication may also be necessary with some children.

Additional information on all disorders and all teaching and treatment methods described in this chapter can be obtained by following the links indicated on pages 44–45.

LINKS TO MORE ON PERVASIVE DEVELOPMENTAL DISORDERS

Background information on all PDDs can be located at:

▶ http://www.medicinenet.com/pervasive_development_disorders/article. htm
▶ http://www.chw.org/display/PPF/DocID/22121/router.asp

More specific information for the syndromes discussed in this chapter are available at the following addresses:

Autism
▶ http://www.ninds.nih.gov/disorders/autism/detail_autism.htm
▶ http://www.autisminfo.org.au/

Asperger syndrome
▶ http://kidshealth.org/parent/medical/brain/asperger.html

Rett syndrome
▶ http://www.nichd.nih.gov/health/topics/Rett_Syndrome.cfm

Childhood disintegrative disorder
▶ http://www.mayoclinic.com/health/childhood-disintegrative-disorder/ DS00801

LINKS TO MORE ON TEACHING AND TREATMENT METHODS

Autism
The following manual provides guidelines for implementing a structured program to target behaviours that are often delayed or absent in the child with autism:
▶ Young, R., Partington, C., & Goren, T. (2009). *SPECTRA: Structured program for early childhood therapists working with autism*. Melbourne: ACER Press.

▶ Discrete Trial Training (DTT) for children with autism: http://www.polyxo. com/discretetrial/

▶ Picture Exchange Communication System (PECS): http://autism. healingthresholds.com/therapy/picture-exchange-communication-s

▶ Diet and medication approaches used with autism: http://www. neurologychannel.com/autism/treatment.shtml

▶ Relationship Development Intervention (RDI) and other approaches for autism: http://www.autismspeaks.org/whattodo/index.php and http:// autism.about.com/od/treatmentoptions/a/RDI.htm

Miller Method
▶ http://www.millermethod.org/pdf/chapter19.pdf
▶ http://www.millermethod.org/

Son-Rise
▶ http://autism.about.com/od/treatmentoptions/a/devtherapies_3.htm
▶ http://www.autismtreatmentcenter.org/contents/about_son-rise/what_is_ the_son-rise_program.php

TEACCH approach
▶ http://www.teacch.com/whatis.html

SCERTS model
▶ http://www.scerts.com/the-scerts-model

Music therapy
▶ http://ezinearticles.com/?The-Benefits-of-Music-Therapy-for- Autism&id=432566

DIR/floortime approach
▶ http://www.autismweb.com/floortime.htm

Pivotal response training
▶ Further details on pivotal response training can be found in the document, How to teach pivotal behaviors to children with autism: A training manual, online at: http://www.users.qwest.net/~tbharris/prt.htm

four

Physical disabilities and associated impairments

▶ A physical disability does not automatically impair a student's ability to learn. While it is true that some students with physical impairments do have significant learning problems, assumptions should never be made about an individual's capacity to learn simply on the basis of a physical disability.

▶ In mainstream schools, problems in accommodating children with physical disabilities (especially those in wheelchairs) are usually more related to difficulties with access to buildings, resources and social networks rather than problems with learning.

▶ In the classroom, many accommodations and adaptations may be required to meet these students' needs.

▶ Assistive technology plays a very important role in increasing the ability of physically disabled students to participate in learning activities.

Physical disabilities are present in a relatively small but diverse group of students in our schools. The intelligence levels for students with physical disabilities cover the full range from giftedness to severe intellectual disability. Disabilities range from some that have little or no influence on learning (because they have not resulted in intellectual or sensory disability) through to other conditions that may involve serious neurological impairments affecting cognition, communication, social development and motor skills.

Students with mild forms of disability, and those with average or above average learning aptitude, will usually attend ordinary schools and will access the mainstream curriculum with extra support. It is essential to ensure that the amount and nature of the support provided does not begin to create unnecessary dependence in the students (Egilson & Traustadottir, 2009). The goal should be to encourage independence.

For students with more severe and complex disabilities, special schools still offer the best placement (Kauffman et al., 2005). In this situation, the curriculum and teaching methods can be more easily adapted to the students' needs, and various forms of therapy can more easily be provided.

It is beyond the scope of this book to provide details of each and every physical disability or health problem. Attention here will be devoted only to some major forms of disability or impairment – cerebral palsy, spina bifida, hydrocephalus, muscular dystrophy, epilepsy, and traumatic brain injury.

Cerebral palsy

Cerebral palsy can be defined as a disorder of movement and posture. It is most often caused by brain injury incurred at the time of birth, but is sometimes associated with prematurity or maternal infection during pregnancy (Pellegrino, 2007). Cerebral palsy has a prevalence rate of approximately 2 to 3 cases per 1000 live births (Hegde, 2008). It is caused by damage or dysfunction in the areas of the brain that control movement. The disorder is not curable, but its negative impact on the individual's physical coordination, mobility, learning capacity and communication skills can be minimised through appropriate intensive therapy, training and education.

Many individuals with cerebral palsy have average or better than aver-age intelligence, but some have a degree of intellectual disability ranging from mild to severe (Turnbull et al., 2007). The disorder exists in several forms (e.g., spasticity, athetosis, ataxia) and at different levels of severity. The type and severity of the condition are related to the particular area or areas of the brain that have been damaged and the extent of that damage.

Spastic cerebral palsy is the most common form of the disorder, found in at least 75 per cent of cases. It is characterised by involuntary muscle contractions that seriously affect bodily control and coordination. Turning the head, for example, can result in an involuntary extension of an arm or leg. Voluntary movement is extremely difficult and limbs may flex or lock into abnormal positions. Even normal actions such as swallowing food can be affected. Some of these students will not be ambulant and will be confined to a wheelchair. *Athetoid cerebral palsy (athetosis)* is characterised not by rigidity of limbs but by uncontrolled tremors and involuntary writhing movements of the body (*dyskinesia*), particularly the arms, neck and head (Hegde, 2008). Most individuals with athetosis are less likely to have intellectual disability but may experience learning problems. *Ataxic cerebral palsy* affects mainly balance, body awareness, posture and gait. *Mixed-type cerebral palsy* combines spasticity with athetoid and ataxic features.

Specific terms are used to denote the extent of paralysis. *Monoplegia* indicates one limb only involved. *Hemiplegia* indicates one side of the body is involved. With *diplegia*, both sides of the body are affected. *Paraplegia* involves both legs. *Quadraplegia* indicates that all four limbs are involved.

The spastic form of cerebral palsy is often accompanied by vision problems, hearing loss, speech and language difficulties, seizures or behavioural or emotional disorders. Major difficulties with eye-muscle control can lead these children to find close-focus tasks such as looking at pictures, attempting to read print or working with numbers physically exhausting and stressful, so they tend to avoid engaging in such activity. Learning problems are often exacerbated by visual perceptual difficulties. Medication taken to control epilepsy can often have the effect of reducing the individual's level of arousal and responsiveness in class, thus adding to problems in learning.

In addition to difficulties with movement and speech, many children with cerebral palsy also tend to tire easily and have difficulty attending to tasks for more than brief periods of time. Those with moderate to severe disability take a very long time to perform even basic physical actions (e.g., pointing to a picture, picking up an object). However, it is important that teachers and aides do not intervene and assist too quickly since it is essential that the student carry through the action. Severely disabled students often require special physical positioning and repositioning in their desks or wheelchairs in order to make best use of their available coordinated movements and to avoid discomfort. They rely on the teacher or an aide to lift and move them.

Some children with cerebral palsy may not develop speech, although their receptive language and understanding may be normal. In their attempts to vocalise they may produce unintelligible sounds and their laughter may be loud and distorted. In addition, they may exhibit other symptoms such as inability to control tongue and face muscles, resulting in drooling and facial contortions. These physical problems are outside the individual's control but they create potential barriers for easy social acceptance and can cause these individuals to be judged as less intelligent than they actually are.

For almost all students with spastic cerebral palsy, intensive communication training is essential (Kuder, 2008) and parents should be actively involved in continuing such training at home (Hegde & Maul, 2006). One of the main priorities for individuals without intelligible speech is to provide an alternative method of communication such as sign language, a communication board or a computer-assisted system (Alant & Lloyd, 2005; Heller & Bigge, 2005).

Treatments and approaches for cerebral palsy

Numerous approaches and interventions have been devised to assist individuals with cerebral palsy to function to the best of their abilities. A visit to any school or centre for students with physical disabilities will reveal most of the treatments and methods described below.

Physiotherapy

All students with cerebral palsy benefit from regular and frequent physical therapy to help them increase their range of movement and to maintain motor skills already acquired (Hegde & Maul, 2006; Miller, 2007). Without such therapy many individuals would become more restricted in their movement over time. Physical therapy uses specific exercises and exercise equipment to help muscles and joints to function more normally and to prevent or correct any deformities (Murray-Leslie & Critchley, 2003).

Occupational therapy

Occupational therapy is concerned with finding ways to help an individual function more independently in daily life. For example, cutlery may be adapted to make eating easier, styrofoam wedges or pads may be devised to help position the person more effectively to carry out certain tasks, or a more suitable wheelchair or walking frame may be designed. The occupational therapist may work closely with experts in assistive technology to help devise appropriate communication or other aids to meet each individual's unique needs. Occupational therapy is also concerned with teaching students with physical disabilities new activities and skills for learning and for recreation.

Surgery

Surgery is sometimes performed on up to 10 per cent of individuals with cerebral palsy. This aims to release constricted muscles and joints and allow the person to assume a more normal posture. Lengthening muscles and tendons can often significantly improve both posture and mobility and can sometimes minimise muscle contractions and spasms. It must be noted, however, that surgery is not always effective.

Assistive technology

Assistive technology plays a major role in the education of students with physical disabilities by enhancing movement, participation and communication, thus facilitating access to the curriculum (Best et al., 2005; Desch, 2007). According to Workinger (2005), the reasons for providing appropriate assistive technology are to compensate for physical or communicative skills that the individual is unable to perform unaided, to increase independent functioning, and to increase social participation. Assistive technology is also invaluable in the rehabilitation of individuals recovering from serious accidents that have resulted in amputations or partial paralysis. The technology includes equipment such as slant-top desks; pencil grips; modified scissor grips; specially designed seating, pads and wedges to help position a child for optimum functioning; electric wheelchairs and walking and standing frames, as well as high-tech adaptations such as modified computer screens and keyboards or switching devices activated by head-pointers or eye movement.

Counselling services

Personal counselling is often needed for students with cerebral palsy who may otherwise develop emotional problems associated with the limiting effects of their disability (Kennedy, 2007). In particular, counsellors may need to address issues related to sexuality, social acceptance and depression.

Neurodevelopmental therapy (NDT)

Underlying most physical intervention methods is a belief that, at a neurological level, effective neural pathways or circuits can be created or recreated by frequent repetition of physical actions, and that these pathways will then bypass or compensate for any damaged areas in the brain (Edwards et al., 2003). Such approaches assume that the brain is still responsive to change and adaptation.

Neurodevelopmental therapy (NDT), also known as the Bobath Method, aims to provide the child with sensorimotor experiences to improve muscle tone and encourage the development of more normal movement patterns (Workinger, 2005). Some of the activities entail handling techniques by caregivers that control or restrict abnormal movement patterns while facilitating more normal responses. This aspect is similar to the approach known as constraint-induced therapy or forced-use therapy, in which a more functional limb is restrained to encourage the use of the less functional limb to bring about improvement (Taub et al., 2004). Teachers, aides and caregivers are trained to integrate these special handling techniques into the daily play activities of the child to provide relevance and to increase frequency and intensity of treatment. Kurtz (2007) comments that although there may be some improvement in the range of motion and quality of movement during the treatment period using this approach, there is no strong evidence of longer-term changes that produce normal movement.

Conductive education

This approach was developed in Hungary by Andras Peto in the 1940s (Hari & Akos, 1988) and is now used in many other countries where training centres have been established. It has also been adopted as the central method in a number of special schools for students with physical disabilities.

Conductive education is based on the belief that through intensive practice of movement patterns, more efficient pathways and connections can be made among the relevant neurons in the motor cortex to overcome or reduce the effects of damage in that area. A key element in the approach is training the child with cerebral palsy to use speech (or inner speech) to initiate and regulate a particular movement. For example, the child may say and count: 'I open my hand, 1 – 2 – 3 – 4. I pick up my book, 1 – 2 – 3 – 4'. The mind consciously directs the muscles to move in the desired action. This self-guided process is termed 'rhythmical intention'. It is applied across a wide range of specifically devised activities (termed 'task series') and is usually taught daily in a small group setting (Edwards et al., 2003). Special items of equipment (e.g., individual benches, ladderback chairs) are essential for carrying out certain of the functional movements. The teacher (called the *conductor*) carries out this training and also integrates speech therapy, occupational therapy and physiotherapy according to individual needs. The approach can be considered 'holistic' in that therapies are provided as an integral part of the overall program, and the activities and tasks cover daily living skills as well as communication and readiness for school learning.

Despite strong support from parents and much anecdotal evidence of effectiveness in individual cases, objective research studies have so far failed to find strong evidence of lasting benefits from conductive education. Having carried out a very thorough review of such studies, Darrah et al. (2004) conclude that although there may be specific children for whom the treatment is very effective, when group data are used for purposes of analysis the overall effectiveness is not proven. At the moment it remains uncertain whether conductive education is more effective than traditional physiotherapy in developing and maintaining motor skills.

Kozijavkin method

This method was developed in the Ukraine and is not known widely outside Europe. The treatment involves the child and parents in two weeks of preparatory work with professionals at a treatment centre, followed by 6 to 8 months of regular work at home. Some aspects of the approach seem to borrow techniques from osteopathy (e.g., manipulation of the spine and joints, massage) together with mobility exercises and equipment to improve coordination, strength and movement patterns. Certain features of the program are controversial and as yet there appears to be no research evidence to support the claims made by the providers. Readers must form their own opinions after checking the information available online (see Links box).

Accommodating and teaching students with cerebral palsy

The education of students with cerebral palsy (and other disabilities affecting movement) needs to focus on providing these individuals with opportunities to access the same range of learning experiences as those available to students without handicaps. This may require adaptations to be made to the environment, to the ways in which these students move (or are moved) around the environment and to the teaching methods and instructional resources used (Best et al., 2005).

It will benefit students in wheelchairs and those moving with walking frames, leg-braces and sticks if classroom furniture is arranged to create wider corridors for movement and to give easier access to facilities and resources. Some students with physical disabilities will need to use modified desks or chairs designed for them by an occupational therapist or technician; others may require padded 'wedges' or other specially constructed cushions to enable them to be placed in the optimum position for work. For some students, adapted devices such as pencil grips and page-turners may be required. It is the teacher's responsibility to ensure that the student makes full use of any prescribed equipment. Allowance may need to be made for large and poorly coordinated handwriting, or alternatively the student may need to use a laptop computer with a modified keyboard to type assignments. Papers may need to be taped firmly to the desktop so that they do not get swept away by an accidental arm movement. In lessons where writing or note-taking is involved, the teacher needs to establish a peer support network, using the assistance of classmates. In such cases it is appropriate to allow the student to use photocopied notes from other students. Sometimes, when speech is not a problem, the teacher can encourage the student to submit an assignment as an audiotape rather than an essay.

The instructional needs of children with cerebral palsy will depend mainly on their level of cognitive ability and the extent of their need for intensive therapy. Students with mild cerebral palsy and normal intelligence may simply be slower at completing assignments and will need more time and encouragement. For those with more severe cognitive and physical disabilities, the teaching principles and direct instructional methods summarised in Chapter 1 will apply.

In many special schools, therapy for speech or movement is most often provided by withdrawing the student from class for 30–40 minute sessions during each day. This has the effect of destroying the continuity of the student's learning experiences in the classroom. To minimise the detrimental effect of this situation, the class teacher needs to ensure that important new work is not introduced when the student is out of the room. It is also helpful when the student returns to class to recap briefly what the class has been doing.

Continuity of learning is also destroyed because some students with cerebral palsy have a high absence rate. This is often due to the need to attend treatment or medical appointments during school hours, or can be the result of recurring health problems. Frequent absence of several days means that the teacher will need to provide the student with work to do at home, and may need to enlist the help of the support teacher to provide some short-term remedial assistance for the student when he or she returns to class.

From the above information, it can be seen that inclusion of students with severe cerebral palsy and intellectual disability in the mainstream is problematic. While a few schools have proved adaptable enough to accommodate these students, the majority of students will need to attend special schools where teaching methods, therapies and staff–student ratios can provide much more effective programs to meet their needs.

Epilepsy

Epilepsy is a fairly frequent additional impairment accompanying cerebral palsy and a number of other disabilities related to organic causes. Epilepsy can also be present as a single condition in an otherwise normally healthy child (Peterson et al., 2003).

Epilepsy is a sudden altered state of consciousness and bodily control due to abnormal electrical discharges within specific areas of the cortex of the brain (Hallahan et al., 2009). The severity of these attacks can range from a very mild loss of awareness lasting a few seconds, through to severe seizures in which the person falls to the ground with convulsions and muscular spasms and loses consciousness for several minutes. Between these two extremes several degrees of epileptic attack can occur.

The mild form of epilepsy, often commencing in the early school years, was previously termed 'petit mal' but is now referred to as *absence seizure.* During an absence seizure, the individual loses contact with reality for a few seconds at a time and is often unaware of what has occurred. This can happen many times throughout the day, thus causing a potential learning difficulty in the school context through loss of attention and continuity of experience. Such children are often misunderstood and are reprimanded for daydreaming and for not attending in class.

The more severe form of epilepsy (previously termed 'grand mal') is now referred to as *tonic-clonic seizure*. Brown (1997, p. 556) clearly describes tonic-clonic seizures in these terms:

> The convulsive seizure itself generally starts with eye deviation upward or to one side, sudden loss of consciousness, and rigidity. This tonic stage lasts for about 30 seconds to 1 minute, during which the individual may stop breathing and bite the tongue. A clonic phase follows with rhythmic jerking of the body, lasting 1 to 3 minutes. The clonic phase is usually followed by lethargy or sleep. Incontinence may occur during this period. Upon recovery, the person typically has no memory of the seizure itself.

In tonic-clonic seizures attacks may occur very rarely, or in more acute cases may occur several times a day. Unlike absence seizures in which there is no advance warning, in this case the individual affected may be able to anticipate an oncoming attack through an internal sensation, termed an 'aura'.

Medication with anti-epileptic drugs is reasonably effective in reducing and controlling seizures in 80 per cent of cases. Teachers or school nurses often have the responsibility for administering the medication to a child during the school day, but most children take their medication before leaving home. The exact dosage and type of drug depends on each individual's response to treatment. Dosage may need to be adjusted many times, and teachers should report any unusual behaviour to the child's parents, as this may be caused by variation in the level of medication. Sometimes the drugs can suppress a child's normal level of arousal in class, resulting in diminished attention, motivation and active participation.

All children with epilepsy should be fully informed about their condition from an early age in order to understand it better and to cope with it effectively. They should feel free to discuss their personal concerns at any time with a teacher or school counsellor. It is also necessary to discuss epilepsy with the child's classmates so that they too understand a seizure when they see one and know what to do and what not to do (McInerney & McInerney, 2006). A few studies (e.g., Lewis & Parsons, 2008) have suggested that some mainstream schools are not particularly accepting and supportive of students with epilepsy. This may be due, in part, to teachers' lack of familiarity with the condition and a lack of confidence in managing seizures. All teachers should receive training in this important area of responsibility.

In managing a seizure in the classroom, the child's convulsions should not be forcibly restrained and no attempt should be made to insert any object between the teeth in the belief that this will stop the person from biting the tongue. When convulsions subside, it is advisable to place the child on one side (in the 'recovery position') to keep airways open. Most seizures last much less than 10 minutes. If the child remains unconscious for 15 minutes or more, outside medical help may be required. Provide a full report of any seizure to the child's parents later.

Some instances of epilepsy evident in the primary school years may disappear by late adolescence or early adulthood (*spontaneous remission*). For others, it may mean a lifetime regime of medication and often some restriction in the range of activities the person participates in.

Spina bifida

The term *spina bifida* means 'split or divided spine'. It is classed as a *neural tube defect* affecting approximately 6 children in every 10 000 births (Liptak, 2007). It is a condition in which, during gestation, certain bones in the spine have not completely formed or joined together to cover and protect the nerves of the spinal cord. This can result in total absence of control of muscles in the lower part of the body.

There are various degrees of severity of spina bifida, the mildest form being *spina bifida occulta* where the defect is not obvious to outside inspection and may have little impact on mobility and learning. The more severe form, *spina bifida cystica*, includes *meningocele* (10 per cent of cases) or the more common form *myelomeningocele*. In this condition, neural tissue is exposed outside the spinal canal at birth and is contained within a cystic swelling on the child's back. The cord is often damaged and bodily functions below this point are seriously disrupted, including the use of lower limbs (Murray-Leslie & Critchley, 2003). The individual may need to use a wheelchair or leg braces and walking-sticks. Control of bladder and bowel function is often impaired, necessitating either the use of a catheter tube to drain the bladder and/or the implementation of a careful diet and toileting routine. The management of incontinence presents the greatest social problem for individuals with spina bifida.

Approximately 3 out of 4 children with *myelomeningocele* have average intelligence, but others have some degree of intellectual disability ranging from mild to severe. In the more severe cases the cognitive impairment is not usually directly related to, or caused by, the defect in the spine but rather to some degree of brain damage due to *hydrocephalus* (see below) or other factors. The degree of physical impairment depends mainly on the location of the defect because this determines which nerves are affected. In general, the higher the position of the lesion on the spine, the greater will be the degree of physical impairment (Loomis, 2007).

Some learning difficulties can be associated with *spina bifida cystica*, including perceptual abilities, attention span, memory, speed of response and fine muscle coordination. Lack of mobility may also reduce the individual's range of social interactions. Some may have had limited opportunities and experiences prior to school and this can add to their difficulty in understanding some aspects of the curriculum. Loomis (2007) suggests that in the early years, many of these students may do quite well because the material is fairly simple and does not require higher-order thinking. However, they begin to have problems once the curriculum content becomes more abstract in upper primary and secondary school. Mathematics and reading comprehension are frequently weak areas.

These students learn best when placed in classrooms where systematic teaching and careful monitoring of progress is accompanied by some degree of differentiation in curriculum content and learning tasks. The amount of assistance provided to individuals will vary according to their different abilities. For those who are also intellectually impaired, the teaching principles set out in Chapter 1 obviously apply. For some, modifications to the physical environment of the classroom will also be needed in order to facilitate wheelchair access and movement.

Hydrocephalus

Hydrocephalus is a condition in which there is impairment of the normal circulation and drainage of cerebrospinal fluid within the skull, resulting in increased intra-cranial pressure and potential brain damage. Treatment for hydrocephalus involves the surgical implantation of a catheter into one ventricle in the brain to drain the excess fluid to the abdominal cavity.

A valve is implanted below the skin behind the ear to prevent any back flow of fluid. If treated early in this way, hydrocephalus will not lead to enlargement of the skull and brain damage. Teachers need to be aware that shunts and valves can become blocked, or the site can become infected. If the child with treated hydrocephalus complains of headache or earache, or if he or she appears feverish and irritable, medical advice should be obtained quickly.

Hydrocephalus can exist occasionally as a single condition in the absence of any other disabilities, but it is most commonly associated as a secondary complicating factor in many cases of spina bifida. Approximately 75 per cent of children with myelomeningocele also have this added complication; and those with hydrocephalus normally have more learning problems than those without (Murray-Leslie & Critchley, 2003). In addition, a child with both spina bifida and hydrocephalus will tend to be hospitalised at regular intervals during his or her school life for such events as replacing shunts and valves, lengthening the catheter as the child grows taller, treating urinary tract infections or controlling respiratory problems. This frequent hospitalisation can significantly interrupt schooling, with complex and sequential subjects such as mathematics being most seriously affected by lost instructional time. Many students in this situation require intensive remedial assistance with their schoolwork.

Muscular dystrophy

Muscular dystrophy is an inherited progressive condition in which muscles become weaker over time, leading to loss of independent function and premature death, usually in the late teens or early adulthood (Farrell, 2006). There are several variations of this condition but the most common is *Duchenne muscular dystrophy*, affecting males exclusively. The prevalence rate is reported to be approximately 1 in 3000 births (Fox, 2003). Intelligence is within the normal range, so no special teaching methods are required unless the student has other developmental problems. However, allowances must be made for the student in school in terms of study demands and work output. Personal counselling and emotional support is strongly recommended to help the individual cope with the knowledge of short life expectancy.

Multiple sclerosis (MS)

Another low-incidence disease that is usually diagnosed in the late teens (but can also be found in a few children below the age of 10 years) is multiple sclerosis (MS) (Brissaud et al., 2001). This condition affects rather more females than males. In this disease, the individual's immune system begins to attack the central nervous system causing a gradual breakdown in normal neurological functioning. The rate of progression of MS is difficult to predict because it varies from individual to individual. Often there are periods of remission, during which there is no further deterioration, followed by relapses when symptoms increase. Deficits begin to occur in physical strength, coordination and visual perception. Usually, cognitive ability declines over time, affecting attention, memory and reasoning (MacAllister et al., 2005). Physical symptoms include loss of sensation in fingers and toes, tremors, and bladder and bowel dysfunction. Some, but not all, individuals with MS experience fairly constant pain.

The underlying causes of MS are not yet well understood, and at the moment there is no cure. However, medication and physiotherapy can help control the symptoms and maintain some degree of mobility and coordination. Depression and mood instability are common secondary symptoms, caused mainly by the detrimental effect that MS can have on the individual's quality of life. Life expectancy is within the normal range, but most persons affected with MS reach old age with significant disabilities.

The prevalence rate for MS is between 1 and 150 in every 100 000. Of these cases, only about 5 per cent are children aged below 15 years. However, this may be an underestimation of the true number of children with the disease because often MS remains undiagnosed until the symptoms become more obvious.

Learning problems can occur for students with MS in school because they may have difficulties sustaining attention, are slower at processing information, are less accurate in handwriting or keyboarding, and may exhibit reduced memory capacity. Their educational progress may also be fragmented by frequent absences from school. As with other conditions described above, these students do need a great deal of emotional support in school, and many require additional remedial teaching to help them keep up with their studies.

Traumatic brain injury (TBI)

Severe head injury is now reported to be the most common cause of acquired disabilities in childhood (Michaud et al., 2007) and the incidence is approximately 2 cases per 1000 in the population. Head injury can lead to chronic academic, behavioural, emotional and interpersonal difficulties. The main periods for such injuries are from birth to 5 years, then again in late adolescence, with many more male cases than female (Hegde, 2008). Causes include falling from a height, sports and recreation accidents, motor vehicle accidents, assaults and child abuse (including the 'shaken baby syndrome'). The injuries are classified as 'open head' wounds, where there is penetration by an object, or 'closed head' injury where there is no penetration.

Typical outcomes from brain injury include seizures, motor impairment, cognitive impairment, memory loss and emotional instability. Impairment is generally in proportion to the severity of the injury, but also depends upon the area of the brain most severely affected (Guilmotte, 1997). In some cases, speech and language functions are impaired and the individual requires extensive speech therapy to regain some degree of communicative ability (Kuder, 2008). Frequent difficulties with word and information retrieval from memory (*anomia*) can slow down the individual's speech and cause great frustration. In terms of cognitive impairment, the individual may now have poor ability to plan ahead and may lack sound judgement. Problem-solving ability and reasoning are weakened and information processing is slower. The rate of learning new information or skills (or re-learning old skills) is also much slower than before. The individual may appear sluggish in motivation and responding. In terms of emotional status, the individual with brain injury may exhibit anxiety, irritability, mood swings, inattention, impulsivity, aggression, social withdrawal and depression. The family members of the individual affected by brain injury can be put under great stress and may need regular help and counselling (Oddy, 2003).

Rehabilitation from brain injury depends on a number of factors, including severity of the injury; quality of rehabilitation treatment; age, personality and motivation of the individual and the availability of family and community support (Guilmotte, 1997). Rapid rates of recovery in the first months are often (but not always) indicative of better final outcomes.

Hardman et al. (2005) state that often students with TBI improve dramatically in the first year following their injury, but after that, progress tends to be much slower.

The duties for a teacher of a student with TBI will depend on the severity of the injury and the rate of recovery and rehabilitation, but they typically include the following:

- keeping instructions clear and simple, and not overloading the student with information or tasks
- finding ways to maximise the individual's attention to learning tasks; for example, by removing distractions, providing cues, limiting the amount of information presented and requiring active responses
- helping to compensate for memory loss by presenting visual cues to aid recall of information; encouraging visual imagery; rehearsing and revising information more often than would be necessary with other learners; teaching self-help strategies such as keeping reminder notes and checking the daily timetable
- assisting the individual to plan ahead, set goals for the day and work towards them
- accommodating the student's poor ability to concentrate and slower rate of working
- working in close collaboration with the school counsellor or psychologist to ensure as far as possible that the student does not become withdrawn and socially isolated.

Other disabling conditions

There are, of course, many other physical conditions and health problems that can affect students. Asthma, for example, is an increasingly common condition diagnosed in school-aged children. In the United States, asthma is the leading chronic illness, the third most common reason for hospitalisation and the single most common cause of absenteeism in individuals below the age of 18 years (American Lung Association, 2008). Asthma is an inflammatory condition of the bronchial airways that causes severe breathing problems, coughing and wheezing. It is thought to be caused by allergies, pollutants in the air and viral infections. Asthmatic children benefit from a certain amount of remedial or 'catch up' teaching to help compensate for their frequent absences from school. In addition, close home–school liaison is useful because teachers can then send appropriate work for the child to do at home.

Cystic fibrosis is another respiratory ailment. The prevalence rate is 1 in 2500 births. Some fifty years ago, very few children with this disorder survived to enter primary school, but advances in treatment and management have extended life expectancy to at least 40 years. Cystic fibrosis is an inherited condition in which mucous secretions in the throat become much thicker than normal and will tend to block air passages and congest the lungs (Peterson et al., 2003). Digestion and sweating processes are also affected. Medication and 'chest therapy' can reduce or control the symptoms of respiratory problems to some extent. Diet has to be controlled carefully to maintain adequate nutrition. Cystic fibrosis has no influence on intelligence level, so most children with the condition are in mainstream schools where they learn through normal methods of instruction. However, they do require certain accommodations to be made, such as allowing the child to drink water frequently and to leave the room to clear mucus when necessary. Other children need to be told that the child's frequent coughing is not due to an infectious disease that they are likely to catch. For other advice see the resources listed in the Links box.

Another medical condition that is detected during childhood is Type 1 diabetes (diabetes mellitus, or insulin dependent diabetes). In Australia, approximately 1 child in every 750 will develop diabetes before the age of 19 (Oberklaid & Kaminsky, 2006). The condition is due to the inability of the pancreas to produce sufficient insulin to control glucose levels in the body. If a high level of glucose is present over a long period it is harmful to the functioning of a number of organs in the body, including kidneys and eyes. Symptoms of diabetes in children include tiredness, excessive thirst, frequent urination, bed wetting, loss of appetite, weight loss, and infections on the skin (Sullivan, 2004). Dangerously high glucose levels can even cause a child to lose consciousness (diabetic coma). Treatment always combines careful diet control and an injection of insulin once or twice each day before meals or via an 'insulin pump' which delivers a continuous supply of basal insulin through a catheter implanted under the skin.

Chronic fatigue syndrome (CFS), a debilitating condition commonly associated with adults, can also be found in children (Sullivan, 2004). The exact cause is still not known, but it seems to occur most often following a viral infection, particularly in the case of individuals who are under stress or pressure (e.g., heavy workload, examinations). The main symptoms are excessive tiredness, headaches, sleep disturbance, depression and loss of drive. There is no cure for CFS at the present time and unfortunately the symptoms may persist in milder forms for several years. Teachers need to appreciate the problems a child is having in concentrating and participating in class, and to offer support and understanding.

Allergies, in one form or another, are causing health problems for an increasing number of children. Specific substances (e.g., food, atmospheric pollutants, medicines, chemicals found in the home, insect bites or stings) can cause a marked allergic reaction in some children. Symptoms vary, but can include sneezing, wheezing, hives (uticaria), and swelling of the throat, tongue and face. The most extreme reaction is referred to as 'anaphylactic shock'. Anaphylactic shock is a life-threatening condition in which the respiratory system goes into spasm, the tongue swells, breathing is very difficult, the pulse rate increases, and blood pressure drops. Emergency medical help should be summoned immediately if a child has this severe reaction. All schools should ensure that their staffs are well aware of the need for urgent response in such cases.

Teachers requiring additional information on students with physical and health impairments are referred to the two pre-eminent text titles in the field: *Teaching individuals with physical or multiple disabilities* (5th ed.) by Best, Heller and Bigge (2005), and *Children with disabilities* (6th ed.) by Batshaw, Pellegrino and Roizen (2007).

LINKS TO MORE ON STUDENTS WITH PHYSICAL DISABILITIES

Cerebral palsy

- http://healthlink.mcw.edu/article/931226359.html
- http://www.cerebralpalsysource.com/Types_of_CP/index.html
- Physical therapy for cerebral palsy: http://treatmentofcerebralpalsy.com/07-physical-therapy.html

▶ Bobath Method: http://www.bobath.org.uk/TheBobathApproach.html
▶ Kozijavkin Method: http://www.reha.lviv.ua/components.0.html and
http://www.cerebralpalsyukraine.com/fileadmin/books/rehabilitation_
game.pdf

Epilepsy

▶ General information: http://www.ninds.nih.gov/disorders/epilepsy/
epilepsy.htm
▶ Advice for parents and teachers: http://www.epilepsysociety.org.uk/
AboutEpilepsy/Epilepsyandyou/Childrenandeducation-1

Spina bifida

▶ http://kidshealth.org/parent/system/ill/spina_bifida.html

Hydrocephalus

▶ A fact sheet on hydrocephalus is available from the National Institute of
Neurological Disorders and Stroke: http://www.ninds.nih.gov/disorders/
hydrocephalus/detail_hydrocephalus.htm

Muscular dystrophy

▶ http://www.ninds.nih.gov/disorders/md/md.htm
▶ http://kidshealth.org/parent/medical/genetic/muscular_dystrophy.html
▶ http://www.brighthub.com/education/special/articles/25193.aspx

Multiple sclerosis

▶ http://www.webmd.com/multiple-sclerosis/ms-in-children

Traumatic brain injury

▶ http://www.familiesandcommunities.sa.gov.au/Default.aspx?tabid=1851

Asthma

▶ The American Lung Association: http://www.lungusa.org/site/c.
dvLUK9O0E/b.22782/
▶ National Asthma Council Australia: http://www.nationalasthma.org.au/

Cystic fibrosis

▶ General information: https://ssl.search.live.com/health/article.aspx?id= articles%2fmc%2fpages%2f1%2fDS00287.html&qu=Cystic+Fibrosis

▶ A teacher's guide to cystic fibrosis (2003) available online at: http:// www.cysticfibrosis.ca/pdf/Teachers_Guide.pdf

Sensory impairment: Hearing and vision

- Hearing and vision impairments are relatively low incidence disabilities that can result in significant learning problems and can impede social development.
- Students with these sensory impairments may be of any ability level and may vary significantly in their degree of disability and their response to special teaching methods.
- Teachers are able to implement many simple adaptations to their teaching approach to accommodate students with these sensory impairments.
- Assistive technology plays a vital role in helping students with sensory impairments access the curriculum.

Students with impaired hearing

Hearing impairment can exist as a single disability, but also occurs as an additional handicap in students with other serious disorders such as cerebral palsy and intellectual disability. Some forms of hearing impairment are present from birth (due to genetic factors or to maternal illness during pregnancy) but other forms are acquired as a result of accidents, illness or adverse environmental factors. Hearing loss present from birth (*pre-lingual deafness*) has much more effect on language acquisition and learning than deafness occurring after speech and language have been developed (Herer et al., 2007).

Impaired hearing creates many problems for learning and for socialisation. Not being able to hear the teacher's words and the words of other children in school can cause tremendous frustration for the students affected (Harigai, 2004; MacFadden & Pittman, 2008). The inability to hear language from an early age not only creates a major problem for developing speech but can also have a negative impact on some aspects of intellectual development. For this reason, a priority goal in the education of all children with impaired hearing is to advance their language skills as much as possible. Any improvement in language will allow each child to make better use of his or her intellectual potential, understand much more of the curriculum and develop socially.

The term *hearing impairment* is used to cover all degrees and types of hearing loss, ranging from profound deafness through to mild loss. Individuals are usually referred to as *deaf* if they are unable to detect speech from others and if their own spoken language is affected. The greater the degree of impairment the less likely it is that the individual will develop normal speech and language and the more likely it is that he or she will need special education services. Those who can hear sounds to some extent and can make reasonable use of their residual hearing are either termed hard-of-hearing or partially hearing.

Hearing impairment is measured in terms of decibels (dB). Zero dB is the point from which people with normal hearing can begin to detect the faintest sounds. Normal conversation is usually carried out at an overall sound level of between 40 and 50dB. Loss of hearing is expressed in terms of the amplification required before the individual can hear sounds and is classified at four levels (Hegde, 2008):

▶ mild loss: 16–40 dB
▶ moderate loss: 41–70 dB
▶ severe loss: 71–90 dB
▶ profound loss: > 90 dB

According to recent figures, the prevalence rate of hearing impairment in the adult population below age 65 is around 16 per cent (Agrawal et al., 2008) with a steady increase in prevalence among the elderly. It is estimated that up to 8 per cent of children may have a mild hearing loss, with an additional 2 to 3 per cent exhibiting moderate to severe loss. While severe and profound losses are usually detected very early in the child's life, mild to moderate losses are often detected only after the child has started school.

The principal characteristics of individuals with hearing loss include:

▶ delayed onset of babbling in early infancy
▶ late achievement of language milestones
▶ very slow acquisition of vocabulary
▶ shorter utterances and frequent non-conventional word order
▶ poor understanding of the speech of others
▶ unusual voice quality
▶ problems with learning to read, write and spell.

Types of hearing loss

There are two main types of hearing impairment, each with different causes and with different responses to intervention and to amplification of sound.

Conductive loss

Conductive hearing loss occurs when sounds are not reaching the middle ear or the inner ear (*cochlea*) because of some physical malformation, blockage or damage to the ear canal or middle ear. Common causes are excessive build-up of wax in the outer ear canal, a ruptured eardrum, abnormality of the ear canal, infections in the middle ear (*otitis media*), or dislocation or damage to the tiny bones of the middle ear. Hearing loss due to middle-ear infection is usually temporary and improves when the infection is treated successfully. If infections are allowed to continue untreated, damage may be done to the middle ear resulting in permanent hearing loss. The use of a hearing aid may significantly help an individual with conductive hearing loss.

Sensori-neural loss

Sensori-neural loss is related to the functioning of the inner ear and the auditory nerve. The most serious forms of hearing impairment are often of this type. Many speech sounds remain undetected, and the sounds that are heard may be distorted. Wearing a hearing aid may not help because amplifying a distorted sound does not make it any clearer. It is reported that in some cases individuals with sensori-neural loss are particularly sensitive to loud noises, perceiving them to be 'painfully' loud (Roeser & Downs, 2004).

Methods of communication

For those who are mildly to moderately hearing impaired it is usual to encourage the use of *oral–aural methods* of communication that rely on listening, speech, and lip reading. For those who are severely to profoundly deaf, speech and hearing are not effective methods of communication and must be supplemented by other methods. These methods include sign language, gesture, cued speech and finger-spelling (*manual methods*). Many deaf or hard-of-hearing individuals appear to benefit most from the *total communication method* that combines features of manual and oral–aural approaches.

Sign language

There are different forms of sign language (e.g., Signed English, Auslan, American Sign Language) sharing characteristics in common but also having some unique features. Deaf children from deaf families will almost certainly have been exposed to and become competent in manual communication even before entering school. There is no strong evidence that early exposure to sign language has a detrimental effect on later oral communication skill development such as speech and lip reading (Meadow, 2005); but sign language remains a controversial issue in the field of deaf education. The use of sign language in everyday life is largely confined to the deaf community and it is not very effective for use in the wider world. However, experts suggest that sign language should be respected as a language in its own right, with its own vocabulary, grammar and semantics, and should be valued and encouraged as an effective mode of communication.

Oral–aural approach

For individuals with mild, moderate to severe hearing loss, the oral–aural approach is advocated. The belief underpinning *oralism* is that to be accepted and to succeed in a hearing world you need to be able to communicate through oral–verbal methods. The approach discourages gesture and sign language and stresses instead the use of any residual hearing the individual may have, supplemented by lip reading and speech training. The feasibility of learning to lip read for many hearing-impaired individuals is often greatly overestimated. Lip reading (or speech reading) is an extremely difficult and inaccurate method of interpreting the communication of others. However, the oral–aural approach remains very popular in deaf education because it enables the child to function at school and in the community.

Total communication approach

The relative popularity of signing versus oral–aural approach ebbs and flows from decade to decade. In response, total communication (also known as *simultaneous communication*) deliberately combines signing and gesture with oral methods to help deaf children comprehend and express ideas and opinions to others. A combination of oral and manual training at an early age appears to foster optimum communicative ability (Meadow, 2005).

Assistive technology for hearing impairment

Hearing aids are of various types, including the typical 'behind the ear' or 'in the ear' aid, and radio frequency (FM) aids. An audiologist assesses the specific needs of the child and a hearing aid is prescribed to suit the individual's sound-loss profile. The aid is adjusted as far as possible to give amplification of the specific frequency of sounds needed by the child. No hearing aid fully compensates for hearing loss, even when carefully tailored to the user's characteristics. The great limitation of the conventional type of hearing aid is that it amplifies all sound, including noise in the environment. The advantage of the radio frequency (FM) aid is that it allows the teacher's voice to be received with minimum interference from environmental noise. The teacher wears a small microphone and the child's hearing aid receives the sounds in the same way that a radio receives a broadcast transmission. The child can be anywhere in the class-room and does not need to be close to or facing the teacher, as with the conventional aid.

For the severe to profoundly deaf individual, a totally different alternative to a hearing aid is the cochlear implant. This device produces the sensation of sound by electrically stimulating the auditory nerve. It has two main internal parts: an array of electrodes implanted inside the cochlear (or inner ear), and a receiver and transmitter embedded in the temporal bone beneath the skin. The external parts, worn behind the ear like a hearing aid, consist of a microphone which detects sound in the environment, a speech processor which filters the sound and processes it into a signal, and a transmitter that sends the signal to the internal receiver. To date, cochlear implants have usually only been done on one side but second or 'bilateral' implants are now becoming more common, either implanted at the same time as the first implant or separately at a later date.

Cochlear implants can be implanted in both children and adults with a functioning auditory nerve. While the patient can begin to perceive the electrical stimulation soon after surgery, it can take some time for people to learn to interpret sounds. Children usually require ongoing speech and language therapy as well as educational support to develop language and communication skills.

The design of implants and speech processors has greatly improved in past decades, and continues to improve, and the scientific literature now generally supports their use because of positive outcomes for language development, speech and literacy. The strong advice and practice of speech pathologists and specialists working with young children with cochlear implants is a oral–aural approach at home and school. Currently, cochlear implants are normally recommended for children who are severe to profoundly deaf. Many developed countries are now carrying out the surgery required to implant this form of assistive device on children as young as six months (Herer et al., 2007).

With advances in technology and intervention, some children who receive cochlear implants at a very young age are achieving near normal speech and language development (Dettman et al., 2007). Although these children may not appear to need any particular special help in the classroom, their hearing is still impaired and they face challenges such as understanding speech in background noise. Young children also need help managing the technology (as with children with hearing aids). In this respect, it is important to maintain good practice in relation to teacher presentation and classroom management.

Teaching hearing-impaired students

Early intervention and active parental involvement are essential elements in language stimulation (Herer et al., 2007). In some cases, speech training and auditory training are advocated for hearing-impaired students. Speech therapists or language teachers may, for example, use speech and articulation coaching based on behavioural principles of modelling, imitation, rewarding and shaping. In recent years, however, speech therapists and teachers have placed even more importance on trying to stimulate language through the use of naturally occurring activities in the classroom, as discussed under language disorders in Chapter 6.

It is recommended that teachers employ the following strategies when working with hearing-impaired students. Many of these points apply equally to working with students with language disorders that are not related directly to hearing loss (see Chapter 6).

- Make sure you involve the student with impaired hearing as much as possible in every activity.
- Aim to increase the student's language skills in every lesson you teach.
- Use clear language when explaining new concepts, and always check for understanding.
- Attract the student's attention when you are about to ask a question or give out information.
- Don't be afraid to introduce new terminology, but teach all new words thoroughly. For example – write new vocabulary on the whiteboard; ensure the student hears the word, sees the word, and says the word; revise new vocabulary regularly; revise new language patterns (e.g., 'Half the size of ...'; 'Combine the ingredients ...'; 'Invert and multiply ...').
- Don't give instructions while there is noise in the classroom.
- Repeat instructions if necessary, while facing the student with impaired hearing.
- Make good use of visual methods of presenting information whenever possible.
- When appropriate, write instructions as short statements on the whiteboard.
- Check frequently that the student is on task and has understood what he or she is required to do.
- When group discussion is taking place, make sure the student with impaired hearing can see the other students who are speaking or answering questions.
- Repeat an answer if you think the student with impaired hearing may not have heard it.
- A student with impaired hearing benefits greatly from seeing your mouth and facial expressions – so don't talk while facing the whiteboard, and do face the class and not the screen when explaining while using an overhead projector.
- Don't walk to the back of the room while giving out important information.

In terms of classroom management and arrangements the following principles apply:

▶ Seat the student where he or she can see you easily, can see the whiteboard, and can observe the other students.

▶ Try to reduce background noise when activities requiring careful listening are conducted.

▶ Don't seat the student with impaired hearing near obvious sources of noise (e.g., fan, air-conditioner, open window, generator).

▶ Ensure that the student has a partner for group or pair activities and assignments.

▶ Encourage other students, when necessary, to assist the student with impaired hearing to complete any work that is set – but without doing all the work for the student.

▶ Where possible, provide the student with printed notes to ensure that key lesson content is available for study, because it may not have been heard during the lesson.

▶ Always check a student's hearing aid very regularly (at least once a day). Often the student will not tell you if a battery is flat or a connection is broken.

▶ Seek advice often from the visiting teacher of the deaf and from the support teacher. Use such advice constructively within your program.

As part of a hearing-impaired student's language development program, careful attention must also be given to the explicit teaching of reading and spelling skills. It is typical of these students that as they progress through primary school they fall three to four years behind the peer group in terms of reading ability (Hardman et al., 2005). This lag in reading has a very detrimental impact on their performance in all subjects across the curriculum. The beginning stages of reading instruction can focus on building a basic sight vocabulary by visual recognition methods, but soon word analysis (decoding) must also be stressed for students with hearing loss (Trezek & Malmgren, 2005). Without knowledge of word structure, the ability of hearing-impaired students to read and spell unfamiliar words will remain seriously deficient. It is also essential when providing reading instruction for hearing-impaired children that due attention be given to extending vocabulary and developing comprehension strategies.

The written expression of deaf children is often reported to be problematic (Antia et al., 2005) with syntax and vocabulary showing major weaknesses. Their difficulties often include inaccurate sentence structure, incorrect verb tenses, difficulties representing plurals correctly and inconsistencies in using correct pronouns. The written work of older deaf students has many of the characteristics of the writing of younger children, and may also contain *deafisms* involving incorrect word order (e.g., 'She got black jeans new', instead of 'She has got new black jeans').

Moores and Martin (2006) cite several studies that support the value of teaching deaf and hard-of-hearing students to apply effective cognitive strategies when tackling classroom learning tasks. Such training normally equips the learner with step-by-step procedures to use when, for example, attempting to get meaning from a textbook chapter, write a report, create a narrative or solve mathematics problems.

Over the past two decades, the movement towards integration and inclusion of children with disabilities into ordinary schools has stressed the value of including hearing-impaired children in the mainstream. It is argued that these children will then experience the optimum level of communication by mixing with other students using normal spoken language. It is believed that regular class placement increases the need and motivation for deaf children to communicate. It is also hoped that students with normal hearing will develop improved understanding and tolerance for individuals who are slightly different from others in the peer group.

Students with impaired vision

The term vision impairment (or the older term *visual impairment*) covers the full spectrum of disability in the domain of sight, from a small loss of vision through to total blindness. The term refers specifically to disorders and conditions that are not corrected merely by wearing spectacles. The impairment may cause general loss of acuity and clarity of vision or it may result in reduction in the field of vision, either central or peripheral. Impaired vision is usually related to conditions in the retina, lens or optic nerve. In the population of students with impaired vision there are those who are deemed totally blind, those who are 'legally' blind, and those with varying degrees of low vision often referred to as partial sight. It is interesting to note that at least 80 per cent of persons classified as legally blind do have some remaining sight (Best, 1992; Davis, 2003). It is estimated that only about 10 per cent of children with vision impairment are totally blind.

Assessing vision

Assessment of children's vision usually involves an investigation of visual acuity, visual field, and visual functioning. Acuity refers to the clarity and 'sharpness' of vision at both near and far distance. Visual field refers to the total span from left to right that an individual can take in when looking straight ahead. Visual functioning refers to the extent to which an individual can make effective use of his or her remaining sight. Measurement of visual acuity does not always provide an adequate indication of visual functioning because individuals with the same measured acuity may make quite different compensations and accommodations for their sight loss. Assessment of visual functioning may include assessment of fine motor coordination, text reading, interpreting pictures or diagrams, spatial awareness, orientation and mobility (Hyvärinen, 2003).

Visual acuity is usually assessed by having the individual identify letters or shapes of decreasing size at a distance of 20 feet (or 6 metres). The smallest letters that can be read provide the measure of acuity, which is expressed in terms such as 20/20 (or 6/6 on the metric scale) meaning normal vision, 20/15 indicating exceptionally good sight, or 20/400 indicating extremely poor sight. A measurement of 6/18 on the metric scale, for example, means that the individual being tested can see at 6 metres what a person with normal vision can see clearly at 18 metres. Vision impairment is often defined as acuity of 20/40 or less in the better eye and/or a visual field of less than 20 degrees diameter. *Low vision* refers to acuity in the range of 20/70 to 20/160 (or 6/18 to 3/60). Individuals with low vision are usually able to use print materials through the use of magnification aids. *Severe impairment* is vision loss in the range of 20/200 to 20/400 (less than 3/60) and many of the people within this category may be classified as legally blind.

Assessment will also determine the partially sighted person's response to illumination. Children with some forms of vision defect are helped greatly by increased illumination (e.g., a reading light placed above their book), but other conditions respond very badly to increased light. Similarly, vision-impaired students differ in their response to magnification, and in their preference for certain background colour, font style and contrasts on a page of text.

In general, the assessments that are made of a student with impaired vision are used to determine the types of adaptations and assistive technology that will help the individual to function most effectively. The general classroom teacher is not expected to interpret detailed quantitative assessment data of this type, but should take advice and guidance from the visiting specialist teacher for vision-impaired services or similar consultant.

Prevalence and impact of impaired vision

In developed countries, vision impairment is a low incidence disability affecting approximately 1 to 2 individuals per 1000 among those below the age of 65 (Tate et al., 2006). Approximately 7 in every 10 000 persons are legally blind. The prevalence rate for impaired vision increases very significantly among the older population. The prevalence rate is also higher in under-developed countries and in communities where health services and infant health care are less readily available.

Vision impairment frequently occurs as a secondary handicapping con- dition in many cases of moderate to severe physical disability (e.g., cerebral palsy, traumatic brain injury) and intellectual disability (Evenhuis et al., 2009). It is reported that at least 30 per cent of individuals with impaired vision have additional needs (Farrell, 2006).

Intelligence levels within the population of vision-impaired persons cover the normal range, and at least 90 per cent of students with low vision (partial sight) are educated in mainstream schools (Erin, 2003). Most students who are blind attend special schools, particularly in the primary years. However, some who are able to compensate effectively for their disability and who have good learning aptitude may transfer to mainstream secondary schools, and later to university (Orsini-Jones, 2009).

The impact of vision loss varies significantly from individual to individual. These differences reflect not only the type and degree of impairment but also the age at which loss of vision occurred. For example, children who are blind from birth are very different in their range of learning experiences and concept development from those who become blind later during childhood or in adult life. Those who had sight for a while have a background of normal visual experiences and images on which to draw.

Bardin and Lewis (2008, p. 474) have commented:

> The very nature of visual impairments can influence the participation of students who are blind or have low vision. Students with visual impairments often miss the subtle, untaught information that provides the basis for understanding key concepts on which general education is based. The resulting gaps in concept development can later affect their ability to infer, predict, comprehend and create during learning activities.

Early development of blind children

Vision is important for developing gross and fine motor skills. It is also essential for acquiring spatial awareness and depth and distance perception. In the case of both blind and low vision children, motor development such as crawling and walking may often be delayed and coordination can be poor (Allen & Cowdrey, 2009). These children benefit from many physical activities that help them develop body awareness and improve movement and coordination.

Cognitive development and concept formation are often delayed by the absence of sight, so early sensory stimulation is vital for very young children with seriously impaired vision and needs to be accompanied by verbal descriptions and explanations (Allen & Cowdrey, 2009; Davis, 2003; Hallahan et al., 2009). Parents and teachers must be aware of the need to provide a blind child with these abundant verbal descriptions. The environment, and events happening within it, must be interpreted through language to increase the child's awareness of things he or she cannot see. Parents, preschool teachers and aides need to use many different activities that encourage the use of the child's intact senses. For example, the child needs to be given different objects to explore through touch. He or she needs to be taught how to examine small objects and large objects in order to build relevant concepts. In addition, the child needs to be encouraged, within the realms of safety, to explore the immediate environment.

In the case of a blind baby, the absence of eye contact between mother and child in the early months can sometimes present an obstacle to normal bonding. Both parent and child need the emotional satisfaction and reinforcement that comes from facial recognition, exchanged smiles and physical contact. Parent guidance usually places a high priority on alerting parents to this potential bonding problem so that they can guard against the tendency to give a blind baby less attention and fewer warm social interactions.

The social development of blind students is often reported to be problematic. This is partly due to lack of opportunity to mix and interact with other children from an early age and observe and acquire social behaviours. It is also due to the fact that blind children can't see the many important non-verbal aspects of social interaction and communication such as nodding in agreement, looking surprised, smiling and respecting personal space when engaging in conversation (Hallahan et al., 2009). Social development is further restricted if members of the peer group feel shy or are lacking in confidence in interacting with a person who is blind. Teachers need to anticipate these problems and be proactive in helping blind students become involved in the social group. For vision-impaired students who are able to cope with learning in the mainstream, their inclusion in regular classes is extremely beneficial for their social development (Davis, 2003).

Types of impairment

Impaired vision has multiple causes. Some vision defects are inherited, including those associated with albinism, congenital cataracts and degeneration of the retina, while others may be due to disease or to medical conditions such as diabetes or tumours. Prematurity and very low birth weight can also contribute to vision problems. Maternal rubella (German measles) during pregnancy is known to cause both vision and hearing impairments in the child.

Among the most frequent diseases or disorders of the eye resulting in impaired vision are:

Glaucoma

A condition caused by a build up of fluid (*aqueous humor*) within the eyeball, resulting in pressure and damage to the retina and optic nerve. Microsurgery is used to drain excess fluid, and early treatment is usually effective in saving sight. Glaucoma is responsible for around 4 to 5 per cent of blindness in children, and a higher percentage in older adults.

Cataracts

This condition affects the lens of the eye, causing clouding and obscuring of vision. Cataracts cause around 15 per cent of blindness in children. Surgery can remove the cataract, but can result in some loss of vision.

Retinitis pigmentosa

This is an inherited condition in which progressive deterioration of the retina occurs. This eventually creates a 'tunnel vision' effect as the areas of the retina responsible for peripheral vision become non-functional. There is no cure for this condition, but diet supplementation (vitamin A) can slow its progress.

Macular degeneration

In children, this inherited condition is often referred to as *macular dystrophy*. An area in the centre of the retina (the macula) becomes non-functional, causing a loss of central field of vision. Assistive technology can help compensate by allowing the individual to make better use of peripheral vision.

Retinopathy of prematurity (ROP)

This is reported to be one of the most common causes of impaired vision in premature or low birth weight children. It is caused by immaturity in the stage of development of the retina when the child is born, but is exacerbated by any abnormalities in the blood supply to the retina. For example, these newborn babies are often given additional oxygen to help them survive for the first few days after birth; but too much oxygen can damage the retina.

Priority needs of students with impaired vision

There are three main areas in which blind children and those with very low vision may need to be taught additional skills. These areas are mobility, orientation and the use of Braille.

Mobility

Individuals with severely impaired vision need mobility skills and confidence to negotiate the outside environment, including crossing roads, catching public transport and locating and using shops. Vision-impaired students need mobility skills to cope with the school environment and to participate actively in appropriate physical activities. Increased mobility adds significantly to the quality of life for persons with impaired vision.

Mobility training is usually regarded as a specialist area of instruction. While the classroom teacher and parents can (and must) assist with the development of mobility skills, a mobility-training expert usually carries out the planning and implementation of the intensive training program. The skills to be taught include:

- *self-protection techniques*: for example, when walking in the school environment, holding the hand and forearm in front of the face for protection while trailing the other hand along the wall or rail; checking for doorways, stairs and obstacles
- *using sound to locate objects*: for example, machinery, air-conditioners, open doorways, traffic noise
- *long-cane skills*: moving about the environment with the aid of a long cane swept lightly on the ground ahead to locate hazards and to check surface textures
- *using electronic travel aids*: for example, spectacles with a sound system built into the frame to warn or inform the person when he or she is nearing an object
- using a person with normal (or adequate) sight as an assistant or guide.

Orientation

Orientation is the term used to indicate that a person with impaired vision is familiar with a particular environment (e.g., the classroom) and at any time knows his or her own position in relation to objects such as furniture, barriers, doors or steps. It will take time for the individual to develop this spatial awareness, but the process can be facilitated if the teacher or aide takes time initially to give the child a conducted 'touch and feel' exploratory tour of each classroom environment that is to be used.

Teachers should realise that for the safety and convenience of students with vision impairment, the physical arrangement of the classroom environment should remain fairly constant and predictable. If furniture has to be moved or some new equipment is introduced into the room, the blind student needs to be informed of that fact and given the opportunity to locate it in relation to other objects. In classrooms it is necessary to avoid placing overhanging obstacles at head height, and to make sure that equipment such as boxes and books are not left on the floor. Doors should not be left half open, with a hard edge projecting into the room.

Mobility and orientation together are two of the primary goals in helping the blind student towards increased independence. Without these skills, the quality of life of the blind person is seriously restricted.

Braille

Braille is of tremendous value as an alternative communication medium for those students who are blind or whose remaining vision does not enable them to perceive enlarged print (Samuels, 2008). Braille is a complex code, so its use with students who are well below average in intelligence is problematic. If a student's cognitive level is such that he or she would experience difficulties in learning to read and write with conventional print, Braille is not going to be an easier code to master. The notion that all vision-impaired students use Braille is entirely false; in fact it is used effectively by only some 4 per cent of vision-impaired children (Farrell, 2006). However, if a blind child's intelligence and tactile perception are adequate, the younger he or she begins to develop some Braille skills the better, because this will enable the child to benefit much more from schooling.

A simplified system similar in principle to Braille is called Moon. It is reported to be easier to learn, particularly for children who have additional disabilities (Davis, 2003; Mednick, 2002; Salisbury, 2008). It uses only 26 raised shapes based on lines and curves to represent the standard alphabet, plus 10 other symbols.

Assistive technology for vision impairment

In the same way that students with physical disabilities or with impaired hearing can be helped to access the curriculum and participate more effectively in daily life through the use of assistive technology, so too children with partial sight can be greatly assisted (Hersh & Johnson, 2008; Salisbury, 2008). Many devices have been designed to enable the student to cope with the medium of print, including magnification aids; closed-circuit television and microfiche readers (both used to enlarge an image); talking calculators; speaking clocks; dictionaries with speech outputs; 'compressed speech' recordings and thermoform duplicators used to reproduce Braille pages or embossed pictures, diagrams and maps (Leventhal, 2008).

Low vision aids are magnification devices or instruments that help the individual with some residual sight to work with maximum visual efficiency. The most recent advances have been in the area of improved computer access. Software is now available to allow the individual to speak his or her input responses and have these converted to text (e.g., *Dragon NaturallySpeaking*) or to have text-to-speech translation (e.g., *ReadPlease*). There are some web browsers that can immediately enlarge on-screen print or can 'read' the content of a web page. Voice recognition systems allow a blind person to give commands and enter data through speech instead of keyboard and mouse.

At a more basic level, some students with impaired vision benefit from modified furniture such as desks with bookstands or angled tops to bring materials closer to the child's eyes without the need to lean over, or with lamp attachments for increased illumination of the page. Some forms of vision impairment respond well to brighter illumination, but in certain conditions very bright light is undesirable. Teachers should obtain advice on illumination and other adaptations from the visiting teacher or from support personnel who are aware of the student's characteristics.

Teaching students with impaired vision

When instructing students who are blind, the teacher's language (together with tactile input) has to compensate for lack of visual information. One only has to think of a subject such as mathematics to appreciate how very difficult it is for a child who has never had sight to learn quantitative concepts and relationships (Osterhaus, 2008). It is even more difficult to master computational skills and meaning of abstract symbols.

Teachers in the mainstream with no prior experience of vision-impaired children may tend to hold fairly low expectations of what these children can accomplish. Inexperienced teachers may not expect enough of a vision-impaired child and may assume that they can't participate in certain activities. It is essential, however, to provide many new challenges for these students and encourage them to do as much as possible (Herold & Dandolo, 2009; Lieberman & Wilson, 2005). Having a problem with vision should not exclude a child from access to the normal curriculum, although often modifications may need to be made.

Simple adaptations that teachers can make include:

▶ seating the student with partial sight in the most advantageous position to be able to see the whiteboard, screen or other display

▶ ensuring that material on the whiteboard or screen is neat and clear, using larger print size than usual

▶ enlarging the print to 24 or even 36 point in all notes and handouts

▶ allowing much more time for students with impaired vision to complete their work

▶ reading the written instructions to students with impaired vision to reduce the amount of time required to begin a task and to ensure that the task is understood

▶ allowing a student with low vision to write and draw with a thicker fibre-tip black ink pen that will produce clear, bold writing

▶ training other students and the classroom aide or assistant to support the student with impaired vision (e.g., by note taking, repeating explanations)

▶ making sure that any assistive equipment is always at hand and in good order.

Additional advice on teaching students with impaired vision can be traced through the resources listed in the Links box. Useful advice can also be found in the texts by Catellano (2005) and Bishop (2004).

Deaf–blind children

Deaf–blind children present a unique challenge to educators because they have impairment in both of the primary sensory modalities normally involved in learning. A blind child's additional hearing handicap is compounded by the fact that he or she cannot see alternative visual forms of communication (sign language, gesture, finger-spelling, lip reading) and cannot interpret non-verbal communication such as the facial expressions of a speaker. The primary method of learning and communicating for deaf–blind students has to be through movement and touch (Hersh & Johnson, 2003) and they require highly specialised teaching (Blaha et al., 2008). Program planning and instructional methods for deaf–blind children have to be tailored to the existing skills and immediate functional needs of each individual child (Marchant, 1992).

Although classified as deaf–blind, some of these children have some residual hearing or sight and it is necessary to capitalise as much as possible on any relative abilities they may have. General strategies used in the early stimulation of children with vision impairment and with deafness (see above) must be applied with these children too, but their rate of response will usually be much slower. Progress will be particularly slow if the individual also has some degree of intellectual disability.

Training in mobility and orientation will be a top priority in programming for a deaf–blind child's curriculum. Without skills in these areas, the individual will remain wholly dependent upon others for meeting even the simplest needs. The child's opportunities to move and explore the environment will also be drastically reduced.

Communication with deaf–blind children must be chiefly through touch. For those children with adequate cognitive ability, modified finger-spelling (where letters are conveyed by touching the child's fingers) can be used. Another option is the deaf–blind manual signing system (also referred to as 'hands-on signing' or 'tactual signing)' where the child places his or her hands on the hands of the communicator to feel the positions (Blaha et al., 2008; Mednick, 2002). Braille or Moon tactile communication systems can also be used with those cognitively able to cope and who have sufficient sensitivity in their finger tips (Hersh & Johnson, 2003).

Most experts on curriculum planning for deaf–blind children (e.g., Haring & Romer, 1995; Huebner et al., 1995) stress four main points:

- the need to assess the individual in great detail to identify the ways in which he or she can respond and can best be stimulated
- the importance of relating the teaching goals to things that will be of immediate value and satisfaction to the child
- the value of carrying out very detailed task analyses in order to work out the simplest possible ways of teaching something new in the absence of sight and hearing
- the need to work collaboratively with other professionals in order to combine expertise from different disciplines.

As with all students who have developmental or acquired disabilities, the ultimate goals for children who are deaf–blind must be to improve as far as possible their quality of life, and to give them a degree of independence and self–determination. Regardless of the degree of impairment, education must address the individual's need to:

▶ access the community
▶ make choices
▶ develop competencies for independent living
▶ be accepted as a member of society
▶ establish positive relationships with other persons.

LINKS TO MORE ON HEARING AND VISION IMPAIRMENTS

▶ Additional background information on hearing impairment is available online at: http://712educators.about.com/gi/dynamic/offsite.htm?zi=1/ XJ/Ya&sdn=712educators&cdn=education&tm=14&gps=352_454_ 1011_474&f=10&tt=14&bt=0&bts=0&zu=http%3A//www.bced.gov. bc.ca/specialed/hearimpair/toc.htm
▶ Several useful information sheets on hearing impairment and teaching approaches from Tasmanian Department of Education are at: http:// www.education.tas.gov.au/school/health/disabilities/disabilitiesinfo/ hearing_impairment
▶ Strategies for teaching students with impaired hearing can be found at: http://www.as.wvu.edu/~scidis/hearing.html
▶ For more advice on blind and vision-impaired students, see: http://www. as.wvu.edu/~scidis/vision.html
▶ Information for classroom assistants and other personnel on teaching vision-impaired students in the mainstream can be found at: http:// www.eshv.org.uk/visionservices/resources/teachersaideadvice.html
▶ For additional strategies for working with vision-impaired students, see: http://learningat.ke7.org.uk/pseweb/New_LSF/Definitions/vs2.html

Language and learning disorders

- A number of learning and communication problems are directly or indirectly associated with language difficulties.
- Speech or language problems may exist as separate areas of difficulty for a student, but these problems are also frequent accompaniments to some forms of intellectual and physical disability.
- Language difficulties underpin many of the problems that some children experience in learning to read, write and spell.
- Specific learning disabilities (SpLD) can be language related or can be due to non-verbal causal factors.

Speech and language disorders

Speech disorders manifest themselves as difficulties in producing speech sounds correctly or delivering spoken words clearly and with appropriate rate and fluency. These problems are often referred to as difficulties with articulation and phonology and may be caused by physical factors, by hearing loss, or by cognitive impairment. *Language disorders* are related to significant difficulties in processing language that is heard (i.e., *receptive language problems*) and/or formulating and expressing one's own ideas or verbal responses clearly and accurately (i.e., *expressive language problems*). Paul (2007, p. 4) states:

Children can be described as having language disorders if they have a significant deficit in learning to talk, understand, or use any aspect of language appropriately relative to both environmental and norm-referenced expectations for children of similar developmental level.

Speech and language problems often exist as single areas of difficulty in otherwise normally developing children, but they also occur quite commonly as secondary difficulties accompanying intellectual disability, hearing impairment and some types of physical handicap (e.g., cerebral palsy) (Hegde & Maul, 2006). Some forms of speech and language difficulty may be due to a delay in development and with assistance the child later overcomes the problem. Other forms are much more deep-seated and require intensive therapy to bring about even modest improvements (Tiegerman-Farber & Radziewicz, 2008).

Speech and language problems can affect several areas in a child's development. For example, Speech Pathology Australia (2006, p. 2) notes that: 'Children with language disorders are more likely to be extremely socially withdrawn, anxious, inhibited or depressed than their age-matched peers with typically developing language skills'. Learning can also be adversely affected in all curriculum areas. Some definitions of language disorder include specific reference to difficulties in comprehending and using written language or other symbol systems. Severe literacy problems affecting both decoding and comprehension are often correlated with general speech and language difficulties (Catts & Kamhi, 2005; Speech Pathology Australia, 2006).

Between 5 per cent and 12 per cent of children have a speech or language disorder that is significant enough to require intervention (Hegde & Maul, 2006; Stuart, 2007) and an additional 7 per cent have minor difficulties (Santrock, 2006). Approximately 1 to 3 per cent of children have a severe form of language disorder that is termed *specific language impairment* (SLI) or *severe language disorder* (SLD).

Specific language impairment (SLI) is not related to intelligence level and is said to exist in children who demonstrate marked receptive and/or expressive language difficulties in the absence of any developmental disability, hearing loss, cognitive deficits, emotional disorders, or social disadvantage (Speech Pathology Australia, 2006; Tiegerman-Farber & Radziewicz, 2008). The child's language skills as measured by standardised tests usually place him or her in the bottom 2 per cent of the population. The causes appear to differ from individual to individual but usually involve significant problems with phonological working memory together with slower and less efficient linguistic processing. Successful comprehension requires a listener to be able to hold information in temporary storage while assimilating and integrating new incoming information. Children with SLI tend to have shorter listening spans than other children, which places them at additional risk in classroom learning situations (Paul, 2007).

Intervention for all forms of speech and language disorder usually involves the language therapist making accurate assessment of the individual's current skills and then using all that is known about normal human language skill acquisition to guide the scope and sequence of an individualised program. Some therapists believe strongly that language intervention should take place in the most natural learning environment and involve authentic activities rather than contrived exercises in a clinical setting (Carr & Firth, 2005). The more natural approach is termed 'milieu teaching' or 'enhanced milieu teaching' depending on the degree of structuring imposed by the therapist. McCauley and Fey (2006, p. 208) define milieu teaching as: 'A conversation-based model of early intervention that uses child interest and initiations as opportunities to model and prompt language use in everyday contexts'. The basic principles are that the teacher–therapist will engage (usually one-to-one) with the child in an activity that requires communication output and listening by the child and immediate responses from the adult. The adult's role is to follow the child's lead as far as possible and to provide modelling, prompting, encouragement, reinforcement, corrective feedback and expansion or shaping of the child's responses as they interact naturally within the context of a task, game or story. Often the focus is on getting the child to say more than he or she otherwise would and to shape these utterances toward improved accuracy, clarity and fluency.

The advantages of a milieu approach are firstly that any improvements in a child's speech and language that occur in a natural situation are more likely to transfer to everyday speech than those achieved in clinical exercises. Secondly, it is an approach that can be taught to parents and caregivers so that it can be extended usefully beyond the school setting (Paul, 2007; McCauley & Fey, 2006). However, Hegde and Maul (2006) suggest that for the most severe cases of disorder, this natural approach is inadequate, particularly if used as the only intervention. They advocate starting intervention with a clinical approach because it can focus on specific goals and the therapist can work systematically towards them, using well-established behavioural techniques of modelling, shaping, repetition and reinforcement. Later, this intensive approach can be supplemented by the more natural milieu approach to facilitate generalisation.

Although children with speech and language disorders clearly need intervention to help minimise or correct their problems, the reality is that it is extremely difficult to bring about improvements. Not only is it impossible in most situations to provide training with the frequency and intensity that is necessary, it is also a matter of motivation on the part of the child. If children have no real desire to change the way they speak and communicate there is little likelihood that they will transfer what is taught in a clinical setting to their everyday lives. This is particularly the case with children who are intellectually disabled.

Central auditory processing disorder (CAPD)

Central auditory processing disorder is closely related to the severe receptive language impairments described above. Auditory processing is a term used to describe what occurs when the brain senses, recognises and interprets incoming sounds. CAPD is said to result in difficulties in recognising or distinguishing between certain individual sounds or words and correctly interpreting and remembering phrases or sentences. For a child in school this difficulty is greatest in a noisy classroom. These children are described as poor listeners and they have difficulty carrying out multi-step instructions or following lengthy explanations. Severe problems with literacy (reading, writing and spelling) can result from this disorder, and CAPD is often diagnosed as a result of referral for a reading difficulty rather than a language problem. The causes of CAPD are not fully understood at this time.

Symptoms for teachers to be aware of include:

- attention problems and distractibility
- difficulty remembering information presented orally
- inability to follow verbal instructions easily
- echolalia (repeating words or phrases without any understanding)
- difficulty recalling names and details

- slowness in processing information and responding
- problems developing vocabulary.

The child with CAPD needs a reasonably quiet learning environment. The teacher must be prepared to repeat and simplify instructions as necessary and check very frequently that the child has understood and is on task. Greater attention also needs to be devoted to the explicit teaching of vocabulary instead of assuming that new words will be picked up incidentally. Direct teaching methods are needed for literacy instruction, with an emphasis on improving the child's attention to sounds within words, learning letter-to-sound relationships (phonics), practising decoding skills and learning effective comprehension strategies.

Specific learning disability (SpLD)

The most comprehensive and widely accepted definition of SpLD comes from legislation in the US where it is stated that:

> The term 'specific learning disability' means a disorder in one or more of the basic psychological processes involved in understanding or in using language, spoken or written, which disorder may manifest itself in imperfect ability to listen, think, speak, write, spell, or to do mathematical calculations. Such term includes such conditions as perceptual disabilities, brain injury, minimal brain dysfunction, dyslexia, and developmental aphasia. Such term does not include a learning problem that is primarily the result of visual, hearing, or motor disabilities; of mental retardation; of emotional disturbance; or of environmental, cultural, or economic disadvantage (US Public Law 108-446, cited in Lerner and Kline 2006, p. 7)

Specific learning disability is the term applied to approximately 3 children in every 100 whose learning difficulties in reading, writing and mathematics cannot be traced to any lack of intelligence, sensory impairment, cultural or linguistic disadvantage or inadequate teaching. The disability manifests itself as a marked discrepancy between intellectual potential and school achievement (APA, 2000). These students exhibit chronic problems in mastering basic academic skills. Some also have problems with social relationships and difficulties with physical skills. They are also extremely slow to respond to remedial intervention, even when this is implemented with frequency and intensity (Kavale, 2005). This slowness to respond or make progress is now regarded as a main marker of a SpLD.

It is often argued that the difficulties of many students with SpLD are not recognised early enough in school, and many of these students are simply judged to be lazy or unmotivated. A few will go on to develop social and emotional problems in response to their lack of school success and will exhibit behaviour difficulties (Hallahan et al., 2009).

The most widely recognised and pervasive SpLD is *dyslexia*. This form of reading problem is thought to be present in approximately 2 per cent of the school population – although some reports place the prevalence rate much higher. Dyslexia is often defined as a disorder causing difficulty in learning to read, despite conventional instruction and adequate intelligence. Dyslexic students have great difficulty in:

▶ understanding and applying phonic decoding principles
▶ building a vocabulary of words recognised by sight
▶ making adequate use of contextual cues to assist word recognition
▶ developing speed and fluency in reading
▶ understanding what has been read.

Other forms of learning disability described in the literature include *dysgraphia* (problems with writing), *dysorthographia* (problems with spelling), *dyscalculia* (problems with number concepts and arithmetic) and *dysnomia* (inability to retrieve words, names, or symbols quickly from memory). Some authorities in the learning disability field tend to attribute these learning problems to neurological deficits or to developmental delay (e.g., Lerner & Kline, 2006). Others suggest that an inefficient learning style is to blame (e.g., Gregory & Chapman, 2002). One particular factor typical of students with a specific reading disability is a lack of awareness of the phonological (speech–sound) aspects of oral language. This difficulty in identifying component sounds within words impairs their ability to master phonic principles and apply decoding strategies for reading and spelling (Muter & Snowling, 2004; Stahl & McKenna, 2006). It is now believed that in the most severe cases of reading disability this poor phonological awareness is often accompanied by a 'naming-speed' deficiency in which the student cannot quickly retrieve a word or a syllable or a letter-sound association from long-term memory (dysnomia). These combined weaknesses create what is termed a 'double deficit' and together make it extremely difficult for the child to develop effective word recognition skills or become a fluent reader (Vukovic & Siegel, 2006). Some children with SpLD also tend to have deficits in working memory, which may lead to difficulties with comprehension of text (Maehler & Schudardt, 2009).

Early intervention is absolutely vital to avoid the detrimental effects of prolonged failure (Yuen et al., 2008). However, the field of learning disability has attracted a number of very unconventional intervention methods or therapies that are of dubious value (Bull, 2009). Parents are often drawn to such approaches and may invest much time and money in a fruitless search for a 'cure' for their child's problems. Research studies, on the other hand, have shown that effective teaching for students with SpLD must involve regular direct instruction and abundant practice in phonics and decoding, together with training in the use of cognitive strategies to aid comprehension. Spelling skills must be taught explicitly alongside phonics, and writing tasks need to be carefully structured to ensure success. Number skills (computation) must be taught directly and then practised to mastery level. Strategies for problem solving must also be directly taught. Remedial teaching in a one-to-one or very small group setting is essential and must be provided frequently to make a real impact. For more detailed information on remedial teaching see Westwood (2007, 2008).

Nonverbal learning disabilities (NLD)

Neuropsychologists have identified a learning disability that is not in any way associated with language skills. The measured verbal IQ of students with this nonverbal learning disability (NLD) is usually very much higher than their nonverbal (performance) IQ. Students with the disorder tend to be quite talkative and to have a good vocabulary (Molenaar-Klumper, 2002). They are also fairly strong in the area of rote memorisation and can retain a great deal of information if it is presented in auditory form. Speaking and listening are their best channels for learning. Their relative fluency in language often leads them to be regarded by their teachers and parents as potentially capable students; so the learning problems that soon become apparent in certain areas are then attributed (wrongly) to laziness or lack of cooperation. The students' main problems are actually associated with gross and fine motor coordination and with spatial awareness. They tend to be clumsy and poorly balanced. Fine motor skills such as handwriting, drawing diagrams or setting down columns of figures in arithmetic appear to be particularly difficult for them, and they are often slow to carry out classroom tasks of this kind. Sometimes they will leave such tasks unfinished. In addition, they seem to have problems with tasks that demand effective use of visual perception, such as interpreting details in pictures, maps, tables or diagrams. It is suggested that such difficulties may stem from some form of subtle dysfunction in the right hemisphere of the brain that controls visuo-spatial, organisational and holistic processing. Their combined coordination and visual perception problems cause them to have difficulties in tasks such as assembling puzzles or models, or when handling equipment in the classroom.

Students with NLD are also relatively weak at interpreting the non-verbal communication of others (facial expressions, gestures, stance, etc.), which then impairs their ability to understand social situations fully. These difficulties can adversely affect their social interactions and their ability to make friends. In this respect, students with NLD share some similarities with students with Asperger syndrome (see Chapter 3), but the two disorders are not the same (Forrest, 2009).

The prevalence of NLD in the school population is not known accurately, partly because it is a relatively new category of learning difficulty and partly because of lack of agreement concerning precise definition and method of identification (Roman, 1998). However, Thompson (1996) suggests that the figure is possibly between 0.1 per cent and 1.0 per cent, and she points out that there are far fewer NLD students than there are students with language-based SpLD.

Students with NLD can be helped significantly once their problem is diagnosed correctly (Martin, 2007). For example, teachers can use much more 'verbal mediation' (i.e., explicit verbal interpretation) when presenting visual and concrete materials. More time can be given for students to complete work and more advice and feedback can be provided. Students can be taught self-regulatory strategies (e.g., self-talk) to enable them to approach classroom tasks systematically and to see them through to completion. The students with very poor handwriting and paperwork can be helped to develop an easier and more legible style. Help is also needed in teaching the student to interpret social situations more accurately and to develop appropriate social skills. An article by Stephanie Morris (2002) provides much helpful information on advancing the social skills of students with NLD (see Links box).

LINKS TO MORE ON LANGUAGE AND LEARNING DISORDERS

Language impairments

▶ Speech Pathology Australia provides an overview of Severe Language Disorder, including issues of identification and intervention from an Australian perspective. Document available online at: http://www.speechpathologyaustralia.org.au/library/Severe%20Lang%20Disorder%20Brief%20Paper%20-%20Jan%2006.pdf

▶ For information on receptive and expressive language disorders see: http://www.betterhealth.vic.gov.au/bhcv2/bhcarticles.nsf/pages/Receptive_language_disorder?Open

Central auditory processing disorder

▶ More information on Central Auditory Processing Disorder (CAPD) is available from the website of the US National Institute on Deafness and Other Communication Disorders (NIDCD) at: http://www.nidcd.nih.gov/health/voice/auditory.asp

Specific learning disability

▶ A general overview of SpLD is available at: http://www.washington.edu/doit/Faculty/Strategies/Disability/LD/

▶ The Australian Psychological Society website provides some additional information of interest to teachers and parents. Available online at: http://www.psychology.org.au/publications/tip_sheets/learning/

Nonverbal learning disability

▶ Morris, S. (2002). Promoting social skills among children with nonverbal learning disabilities. *Teaching Exceptional Children*, *34*, 3, 66–70. Available online at: www.dldcec.org/pdf/Article9.pdf

▶ For additional details on NLD, readers should consult the many excellent papers freely available at NLDline: http://www.nldline.com/

seven

Emotional and behavioural difficulties

KEY ISSUES

▶ Students with emotional and behavioural difficulties present some of the greatest challenges within the school system.

▶ Disturbed students may either externalise their problems in the form of aggression and non-compliance, or they may internalise them in the forms of anxiety, social withdrawal and depression.

▶ Schools need effective counselling services to aid students with emotional and behavioural problems.

▶ Schools also need positive and proactive approaches to behaviour management to prevent troublesome behaviour that disrupts lessons and causes stress for students and for teachers.

There is abundant evidence that behaviour problems and emotional difficulties among schoolchildren are increasing (e.g., Conroy et al., 2009). Several researchers have noted that, relative to other disability groups, students with emotional and behavioural problems experience the worst social, academic and vocational outcomes from schooling (Kern et al., 2009; Nelson et al., 2008). When they leave school, these individuals are also at greater risk for antisocial behaviour and delinquency (Webber & Plotts, 2008). For these reasons, much attention has been devoted over the past decade to early identification of, and intervention for, children's emotional difficulties.

Unlike the disabilities and impairments described in previous chapters, emotional and behavioural disorders (or difficulties) are much less clearly or accurately identified. Indeed, the whole topic of emotional and behavioural disorders contains several contentious issues and there are many conflicting views on what causes such disorders and how they should be treated. According to Spohrer (2008) the term 'emotional and behavioural difficulties' is not a very useful term because it encompasses so many different forms of challenging behaviour exhibited by children for a wide variety of reasons.

Difficulties with definition

Almost all the experts in the field of emotional and behavioural difficulties stress the problem of reaching agreement about the precise meaning of terms such as *emotional disorder, emotional disturbance, maladjustment, behavioural disturbance, behaviour difficulties, conduct disorder, oppositional defiant disorder, personality disorder* and so forth (Danforth & Smith, 2005; Henley et al., 2009; Heward, 2009). Any accurate definition of these terms would require, in the first instance, a clear agreement on what constitutes a 'normal' emotional state, 'normal' behaviour patterns, and 'normal' social adjustment. Reaching such agreement is a difficult if not impossible task. What is regarded as normal or acceptable behaviour in one context or in one culture or social group may be viewed as abnormal and totally unacceptable in another context or culture (Chakraborti-Ghosh, 2008). Adding to the problem of definition is the fact that different professions bring their own perspectives to the field and use their own theoretical models (e.g., medical, behavioural, sociological, psycho-educational) and apply their own terminology when describing or explaining what they see (Hallahan et al., 2009).

What *is* agreed upon is that emotional and behavioural disorders are recognised by their extreme degree of severity, their persistence over a long period of time, and their obvious violation of social or cultural expectations and standards (Kazdin, 1995). It is also agreed that students with emotional and behavioural disorders are usually difficult to teach and difficult to include successfully in mainstream classrooms.

It is reported that some 6 to 10 per cent of school-aged children have emotional and behavioural problems serious enough to warrant special attention (Henley et al., 2009; Heward, 2009). These children display a wide range of difficulties including, on the one hand, externalised symptoms such as hostility, physical and verbal aggression, anti-social behaviour, non-compliance and irritability, and on the other hand internalised symptoms such as social withdrawal, isolation, anxiety and depression. Over the past few decades, the number of students displaying such externalised and internalised problems has steadily increased. Many more boys than girls exhibit the externalised antisocial behaviours described above (Heward, 2009), but there are indications that the number of aggressive and violent girls is increasing. In the past it was believed that girls were more likely to internalise rather than externalise their emotional problems.

Causes of emotional and behavioural difficulties

There are many theories concerning the origins and causes of emotional and behavioural difficulties, ranging from organic factors (biological), genetic predisposition (temperament), traumatic life experiences, dysfunctional family background and other environmental influences. In many cases, the difficulties arise from multiple causes (Hallahan et al., 2009).

Behavioural theory holds that abnormal behaviour and response patterns, whatever the underlying cause, are learned and shaped in the same way that all other behaviours are learned and strengthened over time because they serve some purpose and meet some specific need in the individual's life. For example, some forms of inappropriate behaviour in classroom contexts are reinforced because the student becomes the centre of attention. The teacher pays the student much more attention when he or she is behaving badly than when he or she is working cooperatively. Also, some control techniques used by teachers (e.g., public rebuke or punishment in front of other students) can have the effect of strengthening a student's tough image and status in the peer group. Behavioural theory also suggests that inappropriate and detrimental behaviours can be eliminated and replaced by more appropriate behaviour through intensive use of applied behaviour analysis techniques, cognitive behaviour modification and reinforcement procedures (Webber & Plotts, 2008).

Family background

While it is inappropriate to suggest that a child's problems at school almost always stem from the family environment, in the case of children with challenging behaviour, the evidence is that home background is often very strongly implicated. It is not unusual to find marital discord, psychiatric dysfunction and chronic unemployment in the homes of many children with emotional and behavioural problems. Dysfunctional and aversive family situations are considered a source of antisocial and aggressive models for some children. These families are often characterised by constant negative and coercive verbal exchanges between parents and between other family members, coupled with high rates of anger and abuse. It is reported that a few parents often use quite violent methods of physical punishment, so it is not surprising that the child acquires a belief that aggression is the best or only way to control social situations. Henley et al. (2009, p. 119) have remarked:

> Included among environmental causes are inadequate parenting, over-crowding, racism, lack of employment, overexposure to violence through television and other mass media, peer pressures, and specific social, political and bureaucratic factors that ignore the needs of the young.

School factors

Children who begin school with emotional or behavioural problems may improve in the school environment over time, or they may get worse. Some children begin to show problems for the first time when they arrive at school because inappropriate curriculum and teaching methods, unreasonable expectations, teachers' management styles and daily interaction with other children can all create problems (Heward, 2009). Some students with emotional and behavioural disorders have major difficulties accepting authority and external control, and they have a history of defiant and disruptive behaviour outside the school context. To be successful in school, a student needs to gain control over impulsive and oppositional behaviours and exhibit a more positive, cooperative and compliant attitude toward reasonable authority.

Peer group pressure

Older students in particular may show signs of antisocial or other negative and inappropriate behaviour due to pressure from the peer group. For example, being confrontational with teachers, being disrespectful and uncooperative, ignoring school rules, or making racist remarks may be regarded as the 'norm' for that social group – so, not to behave in that way would cause a member to lose status. The existence of 'gangs' in some schools is another source of peer group pressure and influence. While 'gang' behaviour is extremely undesirable and very disruptive in a school context, it does not necessarily lead to aggression and violence. However, in some circumstances 'socialised aggression' can occur when members of a group acting together harass and bully other students or rival gangs (Gargiulo, 2006).

Describing the symptoms

Worthington and Gargiulo (2006, p. 293) cite the following description of emotional and behavioural disorders, taken from the US *Individuals with Disabilities Education Act* (IDEA):

> The term [emotional and behavioural disorder] means a condition exhibiting one or more of the following characteristics over a period of time and to a marked degree that adversely affects a child's educational performance:
>
> - an inability to learn that cannot be explained by intellectual, sensory or health factors;
> - an inability to build or maintain satisfactory inter-personal relationships with peers and teachers;
> - inappropriate types of behaviour or feelings under normal circumstances;
> - a general pervasive mood of unhappiness or depression;
> - a tendency to develop physical symptoms or fears associated with personal or school problems.

The above description has been criticised by many experts as being too vague and too dependent on subjective judgements to be of much value in accurate identification. However, the five dot-points do help to draw attention to the key features of emotional and behavioural difficulties. If taken together with the following description from the *Special Educational Needs Code of Practice* in the UK, one gets a fairly clear picture of the students in question. The *Code of Practice* (DfES, 2001, Section 7:60) suggests that children and young people who demonstrate features of emotional and behaviour difficulties may be:

▶ disruptive and disturbing
▶ hyperactive and lacking in concentration
▶ immature in social skills
▶ engaging in much challenging behaviour
▶ withdrawn and isolated
▶ displaying additional complex special needs.

In addition, Webber and Plotts (2008) observe that externalised behaviours such as hostility, aggression and bullying put these students at a great risk of social rejection and isolation within the peer group.

The general consensus appears to be that although some students of high ability do have emotional and behavioural difficulties, the majority of students in this category have measured intelligence in the low–average range (IQ 70–100) (Henley et al., 2009; Worthington & Gargiulo, 2006). It must be noted however, that it is not always easy to assess the true intellectual ability of these students since they may not cooperate fully in any testing procedures. Regardless of their true intellectual status, many of these students present as chronic underachievers, with poor results in important areas of the curriculum such as literacy and numeracy (Hallahan et al., 2009; Nelson et al., 2008). Lack of school success adds to their other problems and serves to undermine their self-esteem and motivation, while increasing negativity toward school and teachers (Gargiulo, 2006). It is often far from clear whether the learning problems are a symptom or a primary cause of the child's maladaptive behaviour patterns.

Externalised behavioural problems

Externalised behaviour problems tend to be acted out during the school day as a series of negative social interactions with peers and with teachers. The student concerned is in almost constant conflict and confrontation with those around him or her. There is an active resistance to accepting reasonable discipline and acting in a cooperative manner.

Heward (2009) suggests that *non-compliance* is the most frequent starting point for classroom problems. The teacher may make a simple request which the student ignores or argues against, and the situation then escalates. Non-compliant behaviour results in less task completion, less positive feedback from the teacher, more reprimands, more time spent out of the classroom awaiting counselling or punishment and more emotional confrontations that drain the student's (and teacher's) ability to concentrate on lesson content. Over a period of time, poor achievement, together with increasingly non-compliant and antisocial behaviour, causes the student to become marginalised, alienated and disconnected within the school system (Algozzine & Ysseldyke, 2006). This pattern tends to continue through adolescence, often resulting in dropping out prematurely from school.

Behaviour difficulties and their detrimental outcomes affect the student, the family, the teachers and often the wider school community. It is known that stress levels are increased and health often suffers in parents and teachers who have the day-to-day care of students with disturbed and disturbing behaviour. This is particularly the case with aggressive and disruptive children. Aggression remains perhaps the most serious form of disordered behaviour in a school environment.

Aggression

Aggression clearly falls under the category 'externalised behaviour' and includes verbal aggression (e.g., swearing, threatening, intimidating, arguing), physical aggression toward others (e.g., striking, kicking, bullying), physical aggression toward objects and physical aggression toward teachers. The term 'aggressive behaviour' therefore refers to behaviour that hurts another person or persons indirectly or directly. Children who are aggressive often appear not to develop an awareness or appreciation of the feelings or intentions of others (Henley et al., 2009). Often, aggressive children misinterpret quite neutral social cues as potentially hostile and they react inappropriately to them. For example, in the schoolyard, aggressive children appear to attend selectively to certain cues and interpret them as evidence that their peers are hostile towards them. ('Why did you punch him?' 'He *looked* at me!'). This tendency to make hostile attributions leads these children to respond aggressively to peers who in turn quickly grow to dislike and reject aggressive children for their hostility (Porter, 1996).

Children identified as being very aggressive in the preschool years are often found still to be aggressive at 8 and 16 years of age (Kazdin, 1995). It is also well established that youngsters with aggressive, acting-out behaviour patterns have less chance of good social adjustment and mental health in adult life (Hallahan et al., 2009). The effects of being aggressive and antisocial can start the child or adolescent on a downward path leading to rejection by peers, low self-esteem, depression and delinquency. Unfortunately, the incidence of aggressive and markedly antisocial behaviour is increasing worldwide (Myles & Simpson, 1998).

Some of the risk factors that are predictive of potentially aggressive behaviour in students include:

- a difficult temperament or 'aggressive personality'
- antisocial and non-compliant behaviour detected in the early years
- academic failure and school-related discipline problems
- mental and emotional instability in one or more parent
- inconsistent discipline in the family
- psychological and physical abuse (including sexual abuse)
- dysfunctional family unit
- poor environmental conditions.

There are many different theories concerning the underlying causes of aggression and other behavioural disorders. Some aggressive behaviour may have physical origins (the *biogenic* view), some is learned through imitation of aggressive responses observed in other people (*behavioural* view), some occurs as a result of 'inner conflict' or disturbance in the child's mind (*psychoanalytic* or *psychodynamic* view), some arises when a child fails to achieve acceptance, satisfaction and success (the *humanistic* view) and some when his or her other personal needs are not being met, particularly in the school context (the *psycho-educational* view). It is very rare however, that aggressive behaviour can be traced to one single causal factor. Behaviour usually emerges as the result of a child's innate tendencies or a temperamental predisposition interacting with adverse environmental factors (the *ecological* view). Whatever the underlying causes, aggression quickly becomes part of a person's behavioural repertoire and is shaped and strengthened through normal reinforcement processes. Individuals maintain aggressive responses because they lack non-violent solutions to conflict situations, and they find that aggression works well as a method for establishing control over their environment and gratifying their needs. Where this type of aggressive behaviour has been operating effectively for the child over a considerable period of time it is often extremely difficult to modify.

In the classroom, frustration is frequently an antecedent to aggression. According to Abrams and Segal (1998), sources of frustration for students with potentially aggressive behaviour include:

▶ a disorganised and inconsistent teacher

▶ failure to understand curriculum tasks and activities

▶ lack of success

▶ boredom

▶ absence of positive reinforcement and feedback

▶ irrelevant curriculum

▶ overuse of reprimands and punishment

▶ feelings of powerlessness.

Several specific syndromes of antisocial and aggressive behaviour have been identified, including *conduct disorder* (CD), *oppositional-defiant disorder* (ODD) and *antisocial personality disorder* (APD) or *dissocial personality disorder* (DPD) (APA, 2000). There is considerable overlap among the symptoms of these disorders, with disobedience, disruption to lessons and activities, irritability, lack of anger control and antisocial responses being common to all.

Conduct disorder

Conduct disorder is one of the most common causes for referral of troublesome students to psychological and guidance services. Children and adolescents with conduct disorder exhibit the most extreme cases of non-compliant, aggressive, disruptive and antisocial behaviour. Hyperactivity, impulsiveness, attention deficits, poor school achievement and difficulties in social relationships often accompany conduct disorder. The pattern of behaviour impairs the individual's normal functioning at home and in school, and is usually regarded as 'unmanageable' by parents and teachers (Kazdin, 1995). The prevalence rate of conduct disorder is not clear, but it is suggested that between 2 to 6 per cent of children and youths aged 4 to 18 years may exhibit conduct disorders. In many cases the behaviour problems were evident before the child was 5 years old and appear to be very resistant to most forms of everyday intervention. With early age of onset, the problems tend to continue through adolescence into adult life. Often the child with a conduct disorder will become the youth or adult with antisocial personality disorder (see below). The *Diagnostic and Statistical Manual of Mental Disorders* (APA, 2000) lists a specific pattern of extreme and chronic antisocial behaviours that constitute a conduct disorder. The list includes the following indicators that are evident over a long period of time and occur with high frequency:

- *aggression to people and animals*: bullying, threatening, intimidating, fighting, use of weapons, physical cruelty and (in older subjects) rape
- *destruction of property*: arson, damage, vandalism
- *theft and deceitfulness*: house breaking, car stealing, lying, shoplifting
- *disobedience and violation of rules*: staying out at night, running away from home, truancy.

Oppositional-defiant disorder

Children with oppositional-defiant disorder exhibit a pattern of irritable, negativistic and defiant behaviour toward adults and peers. These children are often viewed by others as spiteful, vindictive and resentful, with a desire not to be controlled by others (Henley et al., 2009). Although they have problems managing their anger, they are less prone to aggression and overt antisocial behaviour than individuals with conduct disorders.

Antisocial personality disorder
(also known as dissocial personality disorder)

A few children, adolescents and adults who are very markedly antisocial, aggressive or volatile and who deliberately inflict pain on others are sometimes diagnosed by psychologists or psychiatrists as having an *antisocial personality disorder*. They appear unable to comprehend the impact that their behaviour has on others, and their behaviour patterns are usually very resistant to change. It is not clear whether the behaviours of individuals exhibiting antisocial personality disorder are due to innate tendencies and temperament or are acquired as the result of learning experiences or emotional deprivation. There are some indications that in certain cases there may be biogenic factors involved.

The *Diagnostic and Statistical Manual of Mental Disorders* (APA, 2000) suggests that antisocial personality disorder begins as a conduct disorder before the age of 15 years, but is soon accompanied by at least four of the following symptoms:

- chronic irresponsible, antisocial, or unlawful behaviour
- irritable and aggressive behaviour (repeated assaults)
- impulsive behaviour and responses
- repeated lying and deception
- reckless disregard for own or others' safety
- callous lack of remorse and lack of empathy.

Approaches to reducing aggressive, disruptive and antisocial behaviour

Teaching aggressive children to be less violent is difficult. Many forms of intervention have been devised ranging from individual, group and family psychotherapy, counselling, social skills training, medication, behaviour modification, behaviour change contracting and cognitive self-control approaches; but none are guaranteed to be effective in all cases. Approaches that use principles from applied behaviour analysis and with the highest degree of structure appear to yield the best results (Webber & Plotts, 2008). The underlying belief in behaviour modification is that problem behaviours must be eliminated and replaced by more acceptable behaviour through a process of modelling and reinforcement. The child is taught alternative ways of responding to stress and frustration other than physical aggression.

Cognitive behaviour modification (CBM) has proved effective with some aggressive and disruptive students. CBM involves teaching students to use 'inner language' ('self talk') to control their own reactions and responses in challenging situations. When CBM is successful with a student, it is a particularly powerful method because it is controlled and maintained by the student, not by outside authority figures such as teachers. In a review of intervention approaches, Kazdin (1995) claims some success for cognitive approaches and cites a number of studies in support. However, a limitation of CBM is that for it to be effective, the student must have a *genuine desire to change* his or her current behaviour pattern – and in many cases that desire is missing. The approach appears to work best with older students who have the maturity and intellectual development to understand the value of controlling their own behaviour.

Devoting class time to discussion and role-playing activities that give all students experience in conflict resolution is strongly advocated by some experts in the field (e.g., Cullinan, 2007; Danforth & Smith, 2005) and is seen as one component of a positive whole-school approach to encourage acceptable behaviour. At an individual level, conflict resolution activities help a student in trouble to reflect upon the situation and to identify ways of bringing about permanent change.

Punishment may have to be included in a behaviour change program; but it must be remembered that punishment is an aversive procedure and while it may stop aggressive behaviour temporarily, it does not, in itself, teach desirable alternative behaviour. Using less aversive approaches such as 'time out' or loss of privileges is important whenever possible.

In terms of preventing or minimising the possibility for aggressive and disruptive outbursts in classrooms, much can be learned from the general proactive tactics used by effective teachers (Hallahan et al., 2009; Nelson et al., 2008). In particular, effective teachers manage their classes well by:

- being alert to everything that is going on in the room
- being able to do two essential things at the same time (instruct and control)
- avoiding dead-spots and poor continuity within a lesson and during transitions
- maintaining a good pace and momentum to the lesson
- taking into account compatibility of student personalities when arranging working groups
- conveying positive expectations of what students will achieve
- giving clear instruction and checking for understanding
- establishing a set of positive classroom rules with known consequences for non-compliance.

Smith (2007) comments that when teachers implement effective teaching procedures, not only does the frequency of problem behaviour reduce but the achievement level of problem students improves as well. Students who are experiencing more success are less likely to engage in disruptive and aggressive behaviour and are easier to control. Heward (2009) cites a number of studies to support the view that inappropriate classroom behaviour can be significantly reduced by effective instructional methods.

Most solutions to problem behaviour will need to come from a collaborative approach involving not only people outside the family (teachers, counsellors, guidance officers, psychologists) but the family members themselves. Kazdin (1995) considers that family therapy and parent management training are among the more positive, comprehensive and enduring ways of dealing with problem behaviour. In both approaches, parents are helped to examine the effectiveness of their own strategies used for interacting with and managing their children. Where necessary, they are taught to apply more positive strategies for dealing with conflict situations in order to avoid constant lose–lose confrontations with the child.

Internalised emotional difficulties

In contrast to the externalised behavioural problems discussed above, some students try to cope with their difficulties by internalising them and keeping them hidden. Examples include extreme anxiety, chronic shyness, social withdrawal and depression. In school situations there is a high probability that some of these problems go unidentified because these students tend not to disrupt lessons and do not create difficulties for teachers and for other students. This does not make the problems any less serious for the individuals concerned. Surprisingly, some recent studies have suggested that some boys, as well as girls, tend to internalise their emotional problems to a serious degree (e.g., McCrae, 2009).

Anxiety

High anxiety is a common characteristic of emotional disturbance, but it also exists in many children who are otherwise well adjusted (Algozzine & Ysseldyke, 2006). It is estimated that some 10 per cent of primary school children have anxiety problems, and these problems can have a major negative impact on both their school achievement and their quality of life (Kendall & Brady, 1995). There is no indication that the adolescent years bring any significant reduction in anxiety, but the focus of anxiety may be quite different for the older student. Adolescence is a time when uncertainty, anxiety, depression and even suicidal thoughts increase for some students.

Anxiety can be defined as alarm, apprehension, tension, or marked uneasiness related to the expectation of pain, embarrassment, discomfort or danger. It may be focused on an object, a person, a situation or an activity; or the anxiety may be unfocused and generalised (Kendall, 1992). The clinical syndrome known as 'generalised anxiety disorder' is identified in a few individuals who spend almost all their waking hours worrying obsessively about a wide range of possible future events and situations (Albano et al., 2003). Anxiety can be so intense and pervasive that it leads to ongoing psychological distress and maladjustment.

In the same way that a predisposition to aggression is thought sometimes to be related to a student's innate temperament, so too anxiety appears to be associated in some cases with a more 'anxious temperament'. Every individual is exposed to stresses in his or her life but the extent to which each person can be resilient and tolerate stress without becoming unduly anxious or depressed appears to vary enormously. Particularly stressful circumstances in life (e.g., an abusive family environment, chronic failure in school, bullying and other forms of victimisation) can be responsible for high and persistent anxiety levels in children.

Anxious children often tend to interpret everyday situations with some degree of distortion. For example, they think that even a simple task is going to be much too difficult and are reluctant to attempt it. Or they anticipate that they are going to be harshly criticised or laughed at when they attempt to do something. In social contexts, they may believe quite wrongly that they are being closely observed and evaluated by others and they fear they will appear deficient or stupid in some way.

A small degree of anxiety helps most people do their best in a learning situation, but more than this minimum amount of anxiety can disrupt learning. In particular, when an individual is in a state of high anxiety the ability to pay attention to and process incoming information is greatly impaired, and working memory capacity is seriously reduced. The phenomenon of 'test anxiety' is, of course, well recognised. Certain individuals reach almost panic level when faced with the prospect of having to take an examination. Fortunately there are self-monitoring strategies that students can learn to help reduce and control their level of anxiety when undertaking tests (Algozzine & Ysseldyke, 2006).

Phobias

A phobia is an emotional state where there is an intense irrational fear of a specific object or situation (e.g., fear of dogs, fear of going to school). When phobic individuals encounter the object or situation, they panic and may display physical symptoms such as feeling sick, difficulty breathing, breaking out in a sweat and increasing heart rate. A phobia goes beyond normal anxiety and can lead to extreme avoidance behaviour. Phobias can develop in any individual of any age or ability level, but they are a frequent occurrence among students with autism.

One intervention used by clinicians for phobic states involves *desensitisation training*. This technique usually requires the individual to be exposed to a series of imagined situations or scenarios ranging from one or two that provoke the least anxiety through to one that provokes high anxiety. These might be displayed as a graded sequence of photographs, or as a role-play situation, or they might be oral descriptions the client or the clinician provides. Treatment begins with the least threatening example and the individual discusses this picture while the clinician gets him or her to relax and to talk about positive feelings. Gradually, over a period of time, the challenge of each new picture or scenario is increased until eventually the individual even feels unthreatened by the most challenging example. Desensitisation can be used in conjunction with *cognitive-behavioural methods* that help individuals explore their own reactions to threatening situations and regulate better their own feelings and responses.

Extreme shyness, withdrawal and social isolation

Teachers are becoming more aware of the significance of extreme shyness and withdrawal as signs of possible emotional disorder. It must be noted, however, that quiet children are not necessarily unhappy or emotionally disturbed – many 'naturally quiet' or solitary children may have no prob-lems at all and no intervention is necessary (Rubin et al., 2003). For example, in families where parents are themselves quiet and reflective rather than loud and assertive, the children may develop similar styles of behaviour. In contrast, intervention is needed with children for whom shyness is overpowering and is impairing their quality of life.

The term 'shy' is very imprecise and subjective. Persons who feel slightly self-conscious or ill at ease in unfamiliar social settings may regard themselves as shy, but might not be judged so by others. A few individuals might be regarded in layman's terms as being 'painfully shy' because their lack of confidence and their frequent embarrassment in social settings are major obstacles to their personal development. Most (probably all) shy individuals wish that they were more confident and assertive.

Individuals vary tremendously in their assertiveness and self-assurance. Many children display shyness at some point in their lives, but most of them quickly develop sufficient confidence to overcome this temporary condition. About 20 per cent of children are thought to be inherently shy as a natural facet of their temperament, and another 20 per cent develop shyness as a result of certain situations experienced in their lives (Kostelnik et al., 2009). Shyness may develop from a loss of self-confidence in social situations due to events in the past that have resulted in embarrassment, humiliation or a feeling of personal inadequacy. The individual begins to withdraw from social interactions (other than those involving family or close friends) and is, for example, fearful of having to participate or perform before a group of people. A preference may develop for solitary pursuits (e.g., watching TV, playing video games, sending e-mails). Internet addiction is a rapidly escalating problem among children of school age. Some of the children most easily affected appear to be those who are shy or are unpopular with peers. Internet addiction (usually regarded as more than 20 hours a week spent at the computer) can lead to serious decline in school performance, health problems, eating disorders and disturbance of sleep patterns (Childnet International, 2006). Voluntary social withdrawal can, at times, be an indication of an emotional problem, so teachers should take time to observe children who appear to be alone most of the time and never wish to join socially with other children. The teacher should seek expert advice if there seems to be cause for concern.

There are two categories of children who spend time in social isolation (Rubin et al., 2003). There are those who would like very much to be part of the social group but for reasons such as extreme shyness, wariness, lack of confidence or social skills or for reasons of rejection by the group are forced to spend most of their time alone along the margins of on-going activities. But there are also those children who appear to prefer their own company rather than constant interaction with others. In terms of need for intervention, the first of these two groups would appear to have priority.

If a child does have a definite problem gaining acceptance in the peer group and is made unhappy by his or her isolation, the teacher does need to intervene. This may apply for example to children with intellectual, physical or sensory disabilities included in mainstream classes where social acceptance can be a problem. Teachers must recognise their responsibility to help these students make friends, join in activities and develop socially. In the literature on socially isolated children, it is usually suggested that careful observation should be made within the classroom and in the schoolyard in order to find the point at which some cautious intervention might be attempted. This might include getting the child to engage in some group play with others, helping the child take a turn in some activity and making sure the child has a partner for a specific activity. The teacher will need to guide, support, prompt and reinforce the child in order to advance the child's socialisation.

Depression

Depression is a frequent accompaniment to anxiety and low-self-esteem (Albano et al., 2003). It seems evident that if an individual is spending most of his or her life in a state of worry and anxiety, depression is likely to develop. Studies have revealed that children at both primary and secondary schools do worry about many issues relating to school, family and life in general (Cheng & Westwood, 2007; Christie & MacMullin, 1998; Tang & Westwood, 2007). For example, during adolescence, failure with peer group relationships and problems with study at school or college can cause a student to become clinically depressed. Depression is thought to affect approximately 3 to 5 per cent of primary school students but up to 15 per cent of teenagers.

'Depression' is a term used rather vaguely in everyday conversation to imply sadness or unhappiness. In clinical and psychological fields, the term is used with greater precision to mean a condition in which the individual not only feels chronically unhappy (*dysphoria*), but also manifests an increasing number of physical and cognitive symptoms such as loss of ability to concentrate, feelings of hopelessness, loss of interest in life, social withdrawal, weakness, disturbed sleeping and eating patterns, loss of weight and somatic physical complaints (Wicks-Nelson & Israel, 2003). Depression may also accompany eating disorders such as *anorexia nervosa* and *bulimia nervosa*, affecting mainly females (Wilson et al., 2003), and is a component of *bipolar disorder* in which the individual exhibits severe mood swings ranging from depression to elation.

The individual with clinical depression usually has to be treated with antidepressant medication and receive regular personal counselling. Long-term intervention must help a person who is prone to anxiety and depression to develop more accurate perceptions of the situations they have found overpowering in the past. They must also be taught coping mechanisms to deal more effectively with life's stresses.

It is only in recent years – partly due to the increasing suicide rate among young people – that schools have acknowledged their responsibility to identify and help children who are depressed. It is essential, particularly in the adolescent years, to be on the look-out for students who appear not to be coping with the personal issues, conflicts and anxieties that accompany the journey toward adulthood. They may be having social acceptance and identity problems within the peer group, there may be problems at home or they could be finding the demands of study too much to manage. Wicks-Nelson and Israel (2003) suggest that one way of helping to prevent the onset of depression is to assist students to develop positive peer relationships and effective social skills. It is also vital that depression-prone and anxious students receive whatever additional tutoring, guidance, learning support and personal counselling they need to minimise the stress associated with study.

Attention deficit hyperactivity disorder (ADHD)

Attention deficit hyperactivity disorder (ADHD) is a syndrome characterised by developmentally inappropriate levels of inattention and distractibility, often accompanied by hyperactivity and impulsivity. ADHD can cause major impairment in everyday functioning at home, at school and in social situations (Glanzman & Blum, 2007; Santrock, 2006). ADHD occurs frequently as an additional complication in children with intellectual disability, autism, acquired brain injury, specific learning disability and emotional disturbance (Heward, 2009). In some children with the disorder, a high level of hyperactivity is not particularly evident – in which case the term attention deficit disorder (ADD) may be used instead.

To be diagnosed as ADHD or ADD the child must exhibit six or more of the nine major symptoms described under 'inattention' and 'hyperactivity' sections in the *Diagnostic and Statistical Manual of Mental Disorders* (APA, 2000). Children with ADHD may also be diagnosed as having *conduct disorders* – defined as a pattern of persistent and repetitive violations of the rights of others or a disregard for age-appropriate social norms and rules (Glanzman & Blum, 2007). Children with both conduct disorders and ADHD tend to display higher levels of physical aggression, more antisocial behaviour, a poor level of school achievement, and higher rates of peer rejection (Allen & Cowdrey, 2009; Hinshaw & Lee, 2003).

In school, children with ADD and ADHD exhibit difficulties in listening, following instructions, completing tasks, remaining seated, managing their own behaviour, making friends and socialising. In general, they display diminished persistence of effort and their teachers often find them difficult to motivate and manage (McInerney & McInerney, 2006). Parents are also stressed by their child's constant level of activity and their problems in learning. ADHD children, while not necessarily below average in intelligence, usually exhibit poor achievement in most school subjects (Lucangeli & Cabrele, 2006).

The overall prevalence rate for ADD/ADHD has been projected to be approximately 8 per cent, with boys far outnumbering girls (Heward, 2009; Scheffler et al., 2009). Wright (2006) suggests that the rate in Australia is between 5 and 9 per cent. However, in the past decade there has been a significant increase in the number of students diagnosed with the disorder, resulting in suggestions that the prevalence rate may be as many as 12 children in every 100. But, it should be noted that the labels ADD and ADHD are often generously misused and applied by parents and teachers to children who are actually displaying a 'normal' range of misbehaviour traits, are bored and/or restless. Accurate assessment of ADHD and ADD can really only be carried out by professionals such as psychologists and paediatricians.

No single cause for ADHD has been identified, although the following have all been put forward as possible explanations: central nervous system dysfunction, subtle forms of brain damage, allergy to specific substances (e.g., food additives), adverse reactions to environmental stimuli (e.g., fluorescent lighting; rotating ceiling fans), inappropriate management of the child at home and maternal alcohol consumption during pregnancy (foetal alcohol syndrome) (Clarren, 2003). It is now generally accepted that the ADHD syndrome may have multiple causes.

The literature indicates that hyperactivity tends to diminish with age even without treatment; but according to Glanzman and Blum (2007) approximately 65 per cent of children diagnosed with ADHD in the early years still have problems during adolescence. In a few cases poor attention, distractibility, impulsiveness and high activity levels persist into adult life.

The belief that ADHD is caused by different factors in different individuals has resulted in different forms of treatment – and what works for one child may not work for another. Treatments have included diet control, medication, psychotherapy, behaviour modification and cognitive behaviour modification. Any approach to the treatment of ADHD needs to attend to *all* factors that may be causing and maintaining the inappropriate behaviours. According to Lerner and Kline (2006) the most effective treatment for ADHD requires the integrated use of effective teaching strategies, a behaviour management plan, parent counselling, a home management program and (in many cases) medication. In the United States, almost 60 per cent of students (4.4 million) with ADHD are taking medication such as Ritalin or Adderal to reduce their hyperactivity and to help them focus attention more effectively on schoolwork (Santrock, 2006). While medication does appear to have a positive impact on students' attention, school achievement and activity level (Scheffler et al., 2009), its use is not without side effects such as loss of appetite, drowsiness, insomnia, headaches and increased nervousness. In Britain and Australia, the use of medication as a first resort is a little less common, and the focus is more on behavioural interventions.

Children with ADHD need structure and predictability in their learning program. They need to be engaged as much as possible in interesting work, at an appropriate level, and in a stable environment (Webber & Plotts, 2008). Enhancing the learning of children with ADHD may also involve:

- providing strong visual input to hold attention
- using computer-assisted learning (CAL)
- teaching the student self-management and organisational skills
- monitoring the child closely during lessons and finding many opportunities to reinforce them when they are on task and productive.

Tourette syndrome

Tourette syndrome is an inherited neuropsychiatric disorder that causes the individual to develop a pattern of vocal and physical tics such as rapid eye blinking, facial grimacing, coughing, throat clearing and snorting. The individual may also involuntarily swear and shout out obscenities. The condition exists in varying degrees from mild (most cases) to severe

(Kostelnik et al., 2009). Tourette syndrome is not related to intelligence, but often these students have learning difficulties and ADHD. The exact prevalence of the syndrome is difficult to determine because some individuals exhibit extremely mild degrees of the disorder. Currently it is believed that between .01 and 0.1 per cent of school children are affected (Lombroso & Scahill, 2008). Some children with ADHD are found to have symptoms of Tourette syndrome along with their generally high activity level (Blum & Mercugliano, 1997). Often the symptoms evident in childhood lessen in severity as the child gets older; but some characteristics remain for life. Causes of this condition are not fully understood, but current evidence suggests that it may be due to abnormal metabolism of the neurotransmitters dopamine and serotonin. Parents with the condition have a 50 per cent chance of passing it on to a child.

Treatment usually involves teaching the individual some self-management techniques to help control the more troublesome tics and behaviours. For extreme cases medication may also be used, particularly if the individual also has ADHD. In situations where a student's Tourette syndrome is not recognised, he or she is often in trouble in school for what is perceived as weird or provocative behaviour. What is actually needed is an understanding and supportive environment.

Interventions

In this chapter, appropriate methods of intervention have been described in each subsection, so it is unnecessary to repeat them here. In general, they range in nature from informal and therapeutic through to highly structured methods based on behavioural principles. In almost all cases, intervention for behavioural and emotional problems can be a very long process and usually requires an effective team approach involving teachers, school counsellors, social workers, peers and family members.

Improving the quality of life for students with these problems should be a very high priority in schools. Helping them to control their own behaviour and helping them become more effectively socialised improves not only their chances in life but reduces the likelihood that as adults they will exhibit antisocial behaviour or suffer from long-term loneliness and depression.

Additional information on all of the above topics can be found at the websites and resources listed below.

LINKS TO MORE ON EMOTIONAL AND BEHAVIOURAL DIFFICULTIES

- General information on a range of emotional and behavioural problems can be located at the United States Department of Health and Human Services at: http://mentalhealth.samhsa.gov/publications/allpubs/CA-0006/default.asp
- Additional advice on management and intervention for behaviour problems is available at: http://specialed.about.com/cs/behaviordisorders/a/Behavior.htm
- Advice on managing aggression in children can be located at: http://www.dupagehealth.org/health_ed/parent_managing.html
- Practical hints for dealing with defiant or non-compliant behaviours is available at Interventions Central website: http://www.jimwrightonline.com/php/interventionista/interventionista_intv_list.php?prob_type=defiance__non_compliance
- Information on anxiety in children can be obtained at: http://www.keepkidshealthy.com/welcome/conditions/Anxiety_Disorders.html
- An ERIC Digest on shyness in children can be found at: http://www.ericdigests.org/pre-928/shy.htm
- Information on depression in children and adolescents is available at: http://www.baltimorepsych.com/cadepress.htm
- http://www.beyondblue.org.au/index.aspx?link_id=4.64
- http://www.kidsmatter.edu.au/
- http://www2.youthbeyondblue.com/ybblue/
- More information on teenage suicide can be located at: http://www.aacap.org/cs/root/facts_for_families/teen_suicide
- For detailed information about internet addiction see Childnet International at: http://www.childnet-int.org/downloads/factsheet_addiction.pdf

Attention deficit hyperactivity disorder

▶ ADHD is described fully at the US National Institute of Mental Health website: http://www.nimh.nih.gov/health/publications/attention-deficit-hyperactivity-disorder/index.shtml

Tourette syndrome

▶ Information is available from the US National Institute of Neurological Disorders and Stroke: http://www.ninds.nih.gov/disorders/tourette/detail_tourette.htm

References

AAMR (American Association on Mental Retardation). (2002). *Mental retardation: Definition, classification, and systems of support* (10th ed.). Washington, DC: AAMR Ad Hoc Committee on Terminology and Classification.

Abrams, B. J., & Segal, A. (1998). How to prevent aggressive behaviour. *Teaching Exceptional Children, 30*(4), 10–15.

Adams, L. (2006). *Group treatment for Asperger syndrome: A social skills curriculum.* San Diego, CA: Plural Publishing.

Adams, L. (2008). *Autism and Asperger syndrome: Busting the myths.* San Diego, CA: Plural Publishing.

Agrawal, Y., Platz, E. A., & Niparko, J. K. (2008). Prevalence of hearing loss and differences by demographic characteristics among US adults. *Archives of Internal Medicine, 168*(14), 1522–1530.

Alant, E. & Lloyd, L. L. (Eds.). (2005). *Augmentative and alternative communication and severe disabilities.* London: Whurr.

Albano, A. M., Chorpita, B. F., & Barlow, D. H. (2003). Childhood anxiety disorders. In E. J. Mash & R. A. Barkley (Eds.), *Child psychopathology* (2nd ed., pp. 279–329). New York: Guilford Press.

Algozzine, R., & Ysseldyke, J. (2006). *Teaching students with emotional disturbance.* Thousand Oaks, CA: Corwin.

Allen, K. E., & Cowdrey, G. E. (2009). *The exceptional child: Inclusion in early childhood education* (6th ed.). Clifton Park, NY: Thomson-Delmar.

Allen, K. E., & Schwartz, I. S. (2000). *The exceptional child: Inclusion in early childhood education* (4th ed.). Albany, NY: Delmar.

American Lung Association. (2008). *Childhood asthma overview.* Retrieved 29 October 2008, from http://www.lungusa.org/site/c.dvLUK9O0E/b.22782/

Antia, S. D., Reed, S., & Kreimeyer, K. H. (2005). Written language of deaf and hard-of-hearing students in public schools. *Journal of Deaf Studies and Deaf Education, 10*(3), 244–55.

APA (American Psychiatric Association). (2000). *Diagnostic and statistical manual of mental disorders: Text Revised (DSM–IV–TR)*. Washington, DC: APA.

Baker, L.,& Welkowitz, L. (2005). *Asperger's Syndrome: Intervening in schools, clinics and communities*. Mahwah, NJ: Erlbaum.

Barbera, M., & Rasmussen, T. (2007). *The verbal behavior approach: How to teach children with autism and related disorders*. London: Jessica Kingsley.

Bardin, J. A., & Lewis, S. (2008). A survey of the academic engagement of students with visual impairments in general education classes. *Journal of Visual Impairment and Blindness, 102*(8), 472–479.

Batshaw, M. L., Shapiro, B., & Farber, M. L. Z. (2007). Developmental delay and intellectual disability. In M. L. Batshaw, L. Pellegrino & N. J. Roizen (Eds.), *Children with disabilities* (6th ed., pp. 245–261). Baltimore, MD: Brookes.

Beirne-Smith, M., Patton, J. R., & Kim, S. H. (2006). *Mental retardation: An introduction to intellectual disabilities* (7th ed.). Upper Saddle River, NJ: Pearson-Merrill-Prentice Hall.

Bell, M. J. (2007). Infections and the fetus. In M. L. Batshaw, L. Pellegrino & N. J. Roizen (Eds.), *Children with disabilities* (6th ed., pp. 71–82). Baltimore, MD: Brookes.

Bennett, S. N. (1999). *Speed of information processing and mental retardation*. BA(Hons) thesis, Department of Psychology, University of Adelaide.

Best, A. R. (1992). *Teaching children with visual impairment*. Milton Keynes: Open University Press.

Best, S. J., Heller, K. W., & Bigge, J. L. (Eds.). (2005). *Teaching individuals with physical or multiple disabilities* (5th ed.). Upper Saddle River, NJ: Pearson-Merrill-Prentice Hall.

Beyer, J., & Gammeltoft, L. (2000). *Autism and play*. London: Jessica Kingsley.

Bishop, V. E. (2004). *Teaching visually impaired children* (3rd ed.). Springfield, IL: Thomas.

Blackbourn, J. M., Patton, J. R., & Trainor, A. (2004). *Exceptional individuals in focus* (7th ed.). Upper Saddle River, NJ: Pearson-Merrill-Prentice Hall.

Blaha, R., Carlson, B., & Moss, K. (2008). *Issues regarding the assessment of vision loss in regard to sign language, finger-spelling, speech reading and cued speech for the student with deafblindness*. Retrieved 14 October 2008, from http://www.tsbvi.edu/Outreach/seehear/archive/sign.html

Blum, N. J. & Mercugliano, M. (1997). Attention-Deficit/Hyperactivity Disorder. In M. L. Batshaw (Ed.), *Children with disabilities* (4th ed., pp. 449–470). Sydney: MacLennan & Petty.

de Boer, S. R. (2007). *How to do discrete trial training.* Austin, TX: ProEd.

Bondy, A., & Frost, L. (2003). Communication strategies for visual learners. In O. I. Lovaas (Ed.), *Teaching individuals with developmental delays* (pp. 291–303). Austin, TX: ProEd.

Braden, M. (2004). The effects of fragile X syndrome on learning. In D. Dew-Hughes (Ed.), *Educating children with fragile X syndrome* (pp. 43–47). London: Routledge-Falmer.

Brissaud, O., Palin, K., Chateil, F., & Pedespan, J. (2001). Multiple sclerosis: pathogenesis and manifestations in children. *Archives de pédiatrie, 8*(9), 969–978.

Browder, D. M., Spooner, F., Ahlgrim-Delzell, L., Harris, A., & Wakeman, S. (2008). A meta-analysis on teaching mathematics to students with significant cognitive disabilities. *Exceptional Children, 74*(4), 407–432.

Brown, I., Percy, M., & Machalek, K. (2007). Education for individuals with intellectual and developmental disabilities. In I. Brown & M. Percy (Eds.), *A comprehensive guide to intellectual and developmental disabilities* (pp. 489–510). Baltimore, MD: Brookes.

Brown, L. W. (1997). Seizure disorders. In M. L. Batshaw (Ed.), *Children with disabilities* (4th ed., pp. 553–593). Sydney: MacLennan & Petty.

Bull, L. (2009). Survey of complementary and alternative therapies used by children with specific learning difficulties (dyslexia). *International Journal of Language and Communication Disorders, 44*(2), 224–235.

Caldwell, P. (2006). *Finding you, finding me: Using Intensive Interaction.* London: Jessica Kingsley.

Cardoso-Martins, C., Peterson, R., Olson, R., & Pennington, B. (2009). Component reading skills in Down syndrome. *Reading and Writing: An Interdisciplinary Journal, 22*(3), 277–292.

Carper, K. G. (2004). Pervasive developmental disorders/ Autism spectrum disorders. In J. M. Blackbourn, J. R. Patton & A. Trainor (Eds.), *Exceptional individuals in focus* (7th ed., pp. 120–135). Upper Saddle River, NJ: Pearson-Merrill-Prentice Hall.

Carr, J. E., & Firth, A. M. (2005). The verbal behavior approach to early and intensive behavioral intervention for autism: A call for additional empirical support. *Journal of Early and Intensive Behavioral Intervention, 2*(1), 18–27.

Cartledge, G. (2005). Learning disabilities and social skills: Reflections. *Learning Disability Quarterly, 28*(2), 179–81.

Cartledge, G., Gardner, R., & Ford, D. Y. (2009). *Diverse learners with exceptionalities.* Upper Saddle River, NJ: Merrill.

Catellano, C. (2005). *Making it work: Educating the blind or visually impaired student in the regular school.* Greenwich, CT: Information Age Publishing.

Catts, H. W., & Kamhi, A. G. (2005). *The connections between language and reading disabilities.* New York: Erlbaum.

Chakraborti-Ghosh, S. (2008). Understanding behaviour disorders: Their perception, acceptance and treatment. A cross-cultural comparison between India and the United States. *International Journal of Special Education, 23*(1), 136–146.

Cheng, N., & Westwood, P. (2007). Self-efficacy, personal worries, and school achievement of primary school students in Hong Kong. *Asia Pacific Education Researcher, 16*(2), 143–154.

Childnet International. (2006*). Internet addiction.* Retrieved 22 June 2009, from http://www.childnet-int.org/downloads/factsheet_addiction.pdf

Chiu, S., Hagerman, R. J., & Leonard, H. (2008). *Pervasive developmental disorders.* Retrieved 17 September 2008, from http://www.emedicine.com/ped/topic1780.htm

Choi, S. H. J., & Nieminen, T. A. (2008). Naturalistic intervention for Asperger syndrome: A case study. *British Journal of Special Education, 35*(2), 85–91.

Christie, E., & MacMullin, C. (1998). What do children worry about? *Australian Journal of Guidance and Counselling, 8*(1), 9–24.

Cipani, E. (2008). *Triumphs in early autism treatment.* New York: Springer.

Clarren, S. K. (2003). Fetal alcohol syndrome. In M. L. Wolraich (Ed.), *Disorders of development and learning* (3rd ed., pp. 235–248). Hamilton, Ontario: Decker.

Cohen, L., & Spenciner, L. J. (2005). *Teaching students with mild and moderate disabilities: Research-based practices.* Upper Saddle River, NJ: Pearson-Merrill-Prentice Hall.

Collins, B. C. (2007). *Moderate and severe disabilities: A foundational approach.* Upper Saddle River, NJ: Pearson-Merrill-Prentice Hall.

Conroy, M. A., Asmus, J. M., Sellers, J. A., & Ladwig, C. N. (2005). The use of an antecedent-based intervention to decrease stereotypic behavior in a general education classroom: A case study. *Focus on Autism and Other Developmental Disabilities, 20*(4), 223–30.

Conroy, M. A., Sutherland, K., Haydon, T., Stormont, M., & Harmon, J. (2009). Preventing and ameliorating young children's chronic problem behaviours: An ecological classroom-based approach. *Psychology in the Schools, 46*(1), 3–17.

Cooper, S. A., Smiley, E., Allan, L. M., Jackson, A., Finlayson, J., Mantry, D., & Morrison, J. (2009). Adults with intellectual disabilities: Prevalence, incidence and remission of self-injurious behaviour and related factors. *Journal of Intellectual Disability Research, 53*(3), 200–216.

Cornish, K. (2004). Cognitive strengths and difficulties. In D. Dew-Hughes (Ed.), *Educating children with fragile X syndrome* (pp. 20–24). London: Routledge-Falmer.

Cronin, M. E., & Patton, J. R. (1993). *Life skills instruction for all students with special needs*. Austin, TX: ProEd.

Crozier, S., & Sileo, N. M. (2005). Encouraging positive behaviour with social stories. *Teaching Exceptional Children, 37*(6), 26–31.

Cullinan, D. (2007). *Students with emotional and behavioral disorders* (2nd ed.). Upper Saddle River, NJ: Pearson-Merrill-Prentice Hall.

Curfs, L., & Frym, J. (1992). Prader-Willi syndrome: A review with special attention to the cognitive and behavioural profile. *Birth Defects: Original Article Series, 28*(1), 99–104.

Danforth, S., & Smith, T. J. (2005). *Engaging troubling students: A constructivist approach*. Thousand Oaks, CA: Corwin.

Darrah J., Watkins, B., Chen, L., & Bonin C. (2004). Effects of conductive education intervention for children with a diagnosis of cerebral palsy: AACPDM Evidence Report. *Developmental Medicine and Child Neurology, 46*, 187–203.

Davis, P. (2003). *Including children with visual impairment in mainstream schools*. London: Fulton.

Davison, P. W., & Myers, G. J. (2007). Environmental toxins. In M. L. Batshaw, L. Pellegrino & N. J. Roizen (Eds.), *Children with disabilities* (6th ed., pp. 61–70). Baltimore, MD: Brookes.

Dempsey, I., & Foreman, P. (2001). A review of educational approaches for individuals with autism. *International Journal of Disability, Development and Education, 48*, 103–116.

Department for Education and Skills (2001). *Special Educational Needs Code of Practice*. Retrieved 13 May 2009, from http://www.teachernet.gov.uk/_doc/3724/SENCodeOfPractice.pdf

Desch, L.W. (2007). Technological assistance. In M. L. Batshaw, L. Pellegrino & N. J. Roizen (Eds.), *Children with disabilities* (6th ed., pp. 557–569). Baltimore, MD: Brookes.

Dettman S. J., Pinder, D., Briggs, R. J., Dowell, R. C., & Leigh, J. R. (2007). Communication development in children who receive the cochlear implant younger than 12 months: Risks versus benefits. *Ear and Hearing, 28*(2), Supplement: 11S–18S.

Dew-Hughes, D. (Ed.). (2004). *Educating children with fragile X syndrome.* London: Routledge-Falmer.

Downing, J. E., & Eichinger, J. (2008). Educating students with diverse strengths and needs together. In J. E. Downing (Ed.), *Including students with severe and multiple disabilities in typical classrooms* (3rd ed., pp. 1–19). Baltimore, MD: Brookes.

Drew, C. J., & Hardman, M. L. (2007). *Intellectual disabilities across the lifespan* (9th ed.). Upper Saddle River, NJ: Pearson-Merrill-Prentice Hall.

Dybdahl, C. S., & Ryan, S. (2009). Inclusion for students with Fetal Alcohol Syndrome: Classroom teachers talk about practice. *Preventing School Failure, 53*(3), 185-196.

Earles-Vollrath, T. L., Cook, K. T., & Ganz, J. B. (2006). *How to develop and implement visual supports.* Austin, TX: ProEd.

Edwards, S., Mawson, S., & Greenwood, R, J. (2003). Physical therapies. In R. J. Greenwood, M. P. Barnes, T. M. McMillan & C. D. Ward (Eds.), *Handbook of neurological rehabilitation* (2nd ed., pp. 179–190). Hove: Psychology Press.

Eggen, P. D., & Kauchak, D. (2007). *Educational psychology: Windows on classrooms* (7th ed.). Upper Saddle River, NJ: Pearson-Merrill-Prentice Hall.

Eggett, A., Old, K., Davidson, L. A., & Howe, C. (2008). *Groupwork for children with autism spectrum disorders ages 11 to 16.* Milton Keynes: Speechmark Publishing.

Egilson, S. T., & Traustadottir, R. (2009). Assistance to pupils with physical disabilities in regular schools: Promoting inclusion or creating dependency? *European Journal of Special Needs Education, 24*(1), 21–36.

Embregts, P. C. M., Didden, R., Huitink, C., & Schrender, N. (2009). Contextual variables affecting aggressive behaviour in individuals with mild to borderline intellectual disabilities who live in a residential facility. *Journal of Intellectual Disability Research, 53*(3), 255–264.

Erin, J. N. (2003). *Educating students with visual impairments.* ERIC Digest E653. Retrieved 16 October 2008, from http://www.hoagiesgifted.org/eric/e653.html

Evenhuis, H. M., Sjoukes, L., Koot, H. M., & Kooijman, A. C. (2009). Does visual impairment lead to additional disability in adults with intellectual disability? *Journal of Intellectual Disability Research, 53*(1), 19–28.

Farrell, M. (2006). *The effective teacher's guide to sensory impairment and physical disability.* Abingdon: Routledge.

Felce, D., Kerr, M., & Hastings, R. P. (2009). A general practice-based study of the relationship between indicators of mental illness and challenging behaviour among adults with intellectual disabilities. *Journal of Intellectual Disability Research*, *53*(3), 243–254.

Firth, G. (2009). A dual aspect model of Intensive Interaction. *British Journal of Learning Disabilities, 37*(1), 43–49.

Fives, C. J. (2008). Vocational assessment of secondary students with disabilities and the school psychologist. *Psychology in the Schools, 45*(6), 508–522.

Forrest, B. (2009). *The boundaries between Asperger and Nonverbal Learning Disability syndromes*. Retrieved 17 June 2009, from http://www.nldline.com/

Fox, M. (2003). *Including children 3–11 with physical disabilities*. London: Fulton.

Frenkel, S., & Bourdin, B. (2009). Verbal, visual and spatio-sequential short-term memory: Assessment of the storage capacity of children and teenagers with Down syndrome. *Journal of Intellectual Disability Research, 53*(2), 152–160.

Frost, L., & Bondy, A. (1994). *The Picture Exchange Communication System: Training manual*. Cherry Hill, NJ: Pyramid Educational Consultants.

Ganz, J. B., Cook, K. T., & Earles-Vollrath, T. L. (2006). *How to write and implement social scripts*. Austin, TX: ProEd.

Gargiulo, R. M. (2006). *Special education in contemporary society* (2nd ed.). Belmont, CA: Wadsworth.

Glanzman, M. M., & Blum, N. J. (2007). Attention deficits and hyperactivity. In M. L. Batshaw, L. Pellegrino & N. J. Roizen (Eds.), *Children with disabilities* (6th ed., pp. 345–365). Baltimore, MD: Brookes.

Gordon, E. S. (2007). Syndromes and inborn errors of metabolism. In M. L. Batshaw, L. Pellegrino & N. J. Roizen (Eds.), *Children with disabilities* (6th ed., pp. 663–697). Baltimore, MD: Brookes.

Greenspan, S. I., & Wieder, S. (2006). *Engaging autism: Using the Floortime Approach to help children relate, communicate and think*. Cambridge, MA: Da Capo Press.

Gregory, G. H., & Chapman, C. (2002). *Differentiated instructional strategies: One size does not fit all*. Thousand Oaks, CA: Corwin.

Gresham, F. M. (2002). Social skills assessment and instruction for students with emotional and behavioral disorders. In K. L. Lane, F. M. Gresham & T. E. O'Shaughnessy (Eds.), *Interventions for children with or at risk for emotional and behavioral disorders* (pp. 242–258). Boston, MA: Allyn and Bacon.

Guilmotte, T. J. (1997). *Pocket guide to brain injury: Cognitive and neurobehavioral rehabilitation*. San Diego, CA: Singular Publishing.

Gutstein, S., Burgess, A., & Montfort, K. (2007). Evaluation of the Relationship Development Intervention Program. *Autism, 11*, 397–412.

Gutstein, S., & Sheely, R. K. (2002). *Relationship Development Intervention with children, adolescents and adults.* London: Jessica Kingsley

Hagerman, R. (2004). Physical and behavioural characteristics of fragile X syndrome. In D. Dew-Hughes (Ed.), *Educating children with fragile X syndrome* (pp. 9–14). London: Routledge-Falmer.

Haggerty, N. K., Black, R. S., & Smith, G. J. (2005). Increasing self-managed coping skills through social stories and apron storytelling. *Teaching Exceptional Children, 37*(4), 40–47.

Hallahan, D. P., Kauffman, J. M., & Pullen, P. C. (2009). *Exceptional learners* (11th ed.). Boston: Allyn & Bacon.

Hardman, M. L., Drew, C. J., & Egan, W. W. (2005). *Human exceptionality: School, community, and family* (8th ed.). Boston: Pearson-Allyn and Bacon.

Hari, M., & Akos, K. (1988). *Conductive education.* London: Routledge.

Harigai, S. (2004). Hearing impairment in school children. In J. Suzuki, T. Kobayashi & K. Koga (Eds.), *Hearing impairment: An invisible disability* (pp. 154–156). Tokyo: Springer-Verlag.

Haring, N. G., & Romer, L. T. (1995). *Welcoming students who are deaf–blind into typical classrooms.* Baltimore, MD: Brookes.

Hegde, M. N. (2008). *Hegde's pocket guide to communication disorders.* Clifton Park, NY: Delmar Learning.

Hegde, M. N., & Maul, C. A. (2006). *Language disorders in children: An evidence-based approach to assessment and treatment.* Boston: Allyn & Bacon.

Heller, K. W., & Bigge, J. L. (2005). Augmentative and alternative communication. In S. J. Best, K. W. Heller & J. L. Bigge (Eds.), *Teaching Individuals with physical or multiple disabilities* (5th ed., pp. 227–274). Upper Saddle River, NJ: Pearson-Merrill-Prentice Hall.

Henley, M., Ramsey, R. S., & Algozzine, R. F. (2009). *Characteristics of and strategies for teaching students with mild disabilities* (6th ed.). Upper Saddle River, NJ: Pearson-Merrill.

Herer, G. R., Knightly, C. A., & Steinberg, A. G. (2007). Hearing: Sounds and silences. In M. L. Batshaw, L. Pellegrino & N. J. Roizen (Eds.), *Children with disabilities* (6th ed., pp. 157–183). Baltimore, MD: Brookes.

Herold, F. & Dandolo, J. (2009). Including visually impaired students in physical education lessons: A case study of teacher and pupil experiences. *British Journal of Visual Impairment, 27*(1), 75–84.

Hersh, M. A., & Johnson, M. A. (2003). *Assistive technology for the hearing-impaired, deaf and deaf–blind.* London: Springer Verlag.

Hersh, M. A., & Johnson, M. A. (2008). *Assistive technology for visually impaired and blind people.* London: Springer.

Heward, W. L. (2009). *Exceptional children* (9th ed.). Upper Saddle River, NJ: Pearson-Merrill-Prentice Hall.

Hines, S., & Bennett, F. (1996). Effectiveness of early intervention for children with Down syndrome. *Mental Retardation and Developmental Disabilities Research Review, 2,* 96–101.

Hinshaw, S. P., & Lee, S. S. (2003). Conduct and oppositional defiant disorders. In E. J. Mash & R. A. Barkley (Eds.), *Child psychopathology* (2nd ed., pp. 144–198). New York: Guilford.

Hodapp, R. M., & Dykens, E. M. (2003). Mental retardation (intellectual disability). In E. J. Mash & R. A. Barkley (Eds.), *Child psychopathology* (2nd ed., pp. 486–519). New York: Guilford.

Holborn, C. S. (2008). Detrimental effects of overestimating the occurrence of autism. *Intellectual and Developmental Disabilities, 46*(3), 243–246.

Holverstott, J. (2005). Promoting self-determination in students. *Intervention in School and Clinic, 41*(1), 39–41.

Howley, M., & Arnold, E. (2005). *Revealing the hidden code: Social stories for people with autism spectrum disorders.* London: Jessica Kingsley.

Huebner, K., Prickett, J., Welch, T., & Joffee, E. (1995). *Hand in hand: Essentials of communication, orientation and mobility for your students who are deaf–blind.* New York: AFB Press.

Humphrey, N. (2008). Including pupils with autistic spectrum disorders in mainstream schools. *Support for Learning, 23*(1), 41–47.

Hutchinson, R., & Kewin, J. (1994). *Sensations and disability: Sensory environments for leisure, Snoezelen, education and therapy.* Exeter: ROMPA.

Hyman, S. L., & Towbin, K. E. (2007). Autism spectrum disorders. In M. L. Batshaw, L. Pellegrino & N. J. Roizen (Eds.), *Children with disabilities* (6th ed., pp. 325–343). Baltimore, MD: Brookes.

Hyvärinen, L. (2003). *Assessment and classification of visual impairment in infants and children.* Paper prepared for the WHO meeting on Classification of Visual Impairment, September 2003. Retrieved 14 October 2008, from: http://www.lea-test.fi/en/assessme/paediatric_low_vis.html

Jameson, J. M., McDonnell, J., Polychronis, S., & Riesen, T. (2008). Embedded constant time delay instruction by peers without disabilities in general education classrooms. *Intellectual and Developmental Disabilities, 46*(5), 346–363.

Joosten, A. V., Bundy, A. C., & Einfeld, S. L. (2009). Intrinsic and extrinsic motivation for stereotypic and repetitive behaviour. *Journal of Autism and Developmental Disorders, 39*(3), 521–531.

Jordan, R. (2001). *Autism with severe learning difficulties.* London: Souvenir Press.

Kauffman, J. M., Landrum, T. J., Mock, D. R., Sayeski, B., & Sayeski, K. L. (2005). Diverse knowledge and skills require a diversity of instructional groups: A position statement. *Remedial and Special Education, 26*(1), 2–6.

Kaufman, B. N. (1995). *Son-Rise: The miracle continues.* Tiburon, CA: Kramer.

Kavale, K. A. (2005). Identifying specific learning disability: Is responsiveness to intervention the answer? *Journal of Learning Disabilities, 38*(6), 553–62.

Kavale, K. A., & Mostert, M. (2004). Social skills interventions for individuals with learning disabilities. *Learning Disability Quarterly, 27*(1), 31–43.

Kazdin, A. E. (1995). *Conduct disorders in childhood and adolescence* (2nd ed.). Thousand Oaks, CA: Sage.

Kendall, P. (Ed.). (1992). *Anxiety disorders in youth: Cognitive behavioural treatments.* Boston: Allyn & Bacon.

Kendall, P., & Brady, E. U. (1995). Comorbidity in the anxiety disorders of childhood. In K. Craig & K. Dobson (Eds.), *Anxiety and depression in adults and children* (pp. 3–36). Thousand Oaks, CA: Sage.

Kennedy, P. (Ed.). (2007). *Psychological management of physical disabilities: A practitioner's guide.* Abingdon: Routledge.

Kern, L., Hilt-Panahon, A., & Sokol, N. G. (2009). Further examining the triangle tip: Investigating support for students with emotional and behavioral needs. *Psychology in the Schools, 46*(1), 18–32.

Kern, P., Wakeford, L., & Aldridge, D. (2007). Improving the performance of young children with autism during self-care tasks using song interventions: A case study. *Music Therapy Perspectives, 25*(1), 43–51.

Kim, J. W., Yoo, H. J., Cho, S. C., Hong, K. E., & Kim, B. N. (2005). Behavioural characteristics of Prader-Willi syndrome in Korea: Comparison with children with mental retardation and normal controls. *Journal of Child Neurology, 20*(2), 134–138.

Kliewer, C. (1998). *Schooling children with Down syndrome*. New York: Teachers College Press.

Koegel, R. L., & Koegel, L. K. (2006). *Pivotal response treatments for autism: Communication, social and academic development*. Baltimore, MD: Brookes.

Kostelnik, M. J., Whiren, A. P., Soderman, A. K., & Gregory, K. M. (2009). *Guiding children's social development and learning* (6th ed.). Clifton Park, NY: Delmar.

Kuder, S. J. (2008). *Teaching students with language and communication disabilities* (3rd ed.). Boston: Pearson-Allyn & Bacon.

Kurtz, L. A. (2007). Physical therapy and occupational therapy. In M. L. Batshaw, L. Pellegrino & N. J. Roizen (Eds.), *Children with disabilities* (6th ed., pp. 61–70). Baltimore, MD: Brookes.

Lerner, J., & Kline, F. (2006). *Learning disabilities and related disorders* (10th ed.). Boston: Houghton Mifflin.

Le Roux, J., Graham, L., & Carrington, S. (1998). Effective teaching for students with Asperger's syndrome in the regular classroom. *Australasian Journal of Special Education, 22*(2), 122–128.

Leventhal, J. (2008). Advice on classroom reading for a child with low vision. *Journal of Visual Impairment and Blindness, 102*(1), 47–49.

Levy, S. E., & Hyman, S. L. (2005). Novel treatments for autistic spectrum disorders. *Mental Retardation and Developmental Disabilities Research Reviews, 11*(2), 131–42.

Lewis, A., & Parsons, S. (2008). Understanding of epilepsy by children and young people with epilepsy. *European Journal of Special Needs Education, 23*(4), 321–335.

Lieberman, L. J., & Wilson, S. (2005). Effects of a sports camp practicum on attitudes toward children with visual impairments and deaf–blindness. *Re:View, 36*(4), 141–53.

Liptak, G. S. (2007). Neural tube defects. In M. L. Batshaw, L. Pellegrino & N. J. Roizen (Eds.), *Children with disabilities* (6th ed., pp. 419–438). Baltimore, MD: Brookes.

Lombroso, P. J., & Scahill, L. (2008). Tourette syndrome and obsessive-compulsive disorder. *Brain Development, 30*(4), 231–237.

Loomis, J. W. (2007). The school-age child: Academic issues. In S. G. Openheimer (Ed.), *Neural tube defects* (pp. 37–60). New York: Informa-Healthcare.

Lovaas, O. I. (1993). The development of a treatment-research project for developmentally disabled and autistic children. *Journal of Applied Behavior Analysis, 26*, 617–630.

Lovaas, O. I., & Smith, T. (2003). Early and intensive behavioral interventions in autism. In A. E. Kazdin & J. R. Weisz (Eds.), *Evidence-based psychotherapies for children and adolescents* (pp. 325–340). New York: Guilford.

Lovering, J. S., & Percy, M. (2007) Down syndrome. In I. Brown & M. Percy (Eds.), *A comprehensive guide to intellectual and developmental disabilities* (pp. 149–172). Baltimore, MD: Brookes.

Lucangeli, D., & Cabrele, S. (2006). Mathematical difficulties and ADHD. *Exceptionality, 14*(1), 53–62.

MacAllister, W. S., & Associates (2005). Cognitive functioning in children and adolescents with multiple sclerosis. *Neurology, 64*(8), 1422–1425.

MacFadden, B., & Pittman, A. (2008). Effect of minimal hearing loss on children's ability to multitask in quiet and in noise. *Language, Speech and Hearing Services in Schools, 39*(3), 342–351.

Macintyre, C. (2002). *Play for children with special needs.* London: Fulton.

MacNeil, B. M., Lopes, V. A., & Minnes, P. M. (2009). Anxiety in children and adolescents with autism spectrum disorders. *Research in Autism Spectrum Disorders, 3*(1), 1–21.

Maehler, C., & Schudardt, K. (2009). Working memory functioning in children with learning disabilities: Does intelligence make a difference? *Journal of Intellectual Disability Research, 53*(1), 3–10.

Marchant, J. M. (1992). Deaf–blind handicapping conditions. In P. McLaughlin & P. Wehman (Eds.), *Developmental disabilities* (pp. 113–123). Boston: Andover Medical.

Martin, M. (2007). *Helping children with nonverbal learning disabilities to flourish.* London: Jessica Kingsley.

McCauley, R. J., & Fey, M. E. (2006). *Treatment of language disorders in children.* Baltimore, MD: Brookes.

McConnell, K., & Ryser, G. R. (2005). *Practical ideas that really work for students with Asperger syndrome.* Austin, TX: ProEd.

McCrae, J. S. (2009). Emotional and behavioral problems reported in Child Welfare over 3 years. *Journal of Emotional and Behavioral Disorders, 17*(1), 17–28.

McInerney, D., & McInerney, V. (2006). *Educational psychology: Constructing learning* (4th ed.). Frenchs Forest, NSW: Prentice Hall.

Meadow, K. P. (2005). Early manual communication in relation to the deaf child's intellectual, social and communicative functioning. *Journal of Deaf Studies and Deaf Education, 10*(4), 321–29.

Mednick, M. (2002). *Supporting children with multiple disabilities.* Birmingham: Questions Publishing Co.

Meilleur, A. S., & Fombonne, E. (2009). Regression of language and non-language skills in Pervasive Developmental Disorders. *Journal of Intellectual Disability Research, 53*(2), 115–124.

Mesibov, G. B., Shea, V., Schopler, E., & Associates. (2005). *The TEACCH approach to autism spectrum disorders.* New York: Kluwer Academic-Plenum.

Michaud, L. J., Duhaime, A. C., Wade, S., Rabin, J. P., Jones, D. O., & Lazar, M. F. (2007). Traumatic brain injury. In M. L. Batshaw, L. Pellegrino & N. J. Roizen (Eds.), *Children with disabilities* (6th ed., pp. 461–476). Baltimore, MD: Brookes.

Miller, A., & Chretien, K. (2007). *The Miller Method: developing the capacities of children on the autism spectrum.* London: Jessica Kingsley.

Miller, F. (2007). *Physical therapy for cerebral palsy.* New York: Springer.

Mitchell, D. (2008). *What really works in special and inclusive education: Using evidence-based teaching strategies.* Abingdon: Routledge.

Molenaar-Klumper, M. (2002). *Nonverbal learning disabilities.* London: Jessica Kingsley.

Moores, D. F., & Martin, D. S. (2006). *Deaf learners: Development in curriculum and instruction.* Washington, DC: Gallaudet University Press.

Morris, S. (2002). Promoting social skills among children with nonverbal learning disabilities. *Teaching Exceptional Children, 34*, 3, 66–70.

Murray-Leslie, C., & Critchley, P. (2003). The young adult with neurological disabilities, with particular reference to cerebral palsy and spina bifida. In R. J. Greenwood, M. P. Barnes, T. M. McMillan & C. D. Ward (Eds.), *Handbook of neurological rehabilitation* (2nd ed., pp. 577–593). Hove: Psychology Press.

Muter, V., & Snowling, C. (2004). *Early reading development and dyslexia.* London: Whurr.

Myles, B., & Simpson, R. (1998). Aggression and violence by school-age children and youth. *Intervention in School and Clinic, 33*(5), 259–264.

Myrbakk, E., & von Tetzchner, S. (2008). Psychiatric disorders and behaviour problems in people with intellectual disability. *Research in Developmental Disabilities: A Multidisciplinary Journal, 29*(4), 316–332.

Neely-Barnes, S., Marcenko, M., & Weber, L. (2008). Does choice influence quality of life for people with mild intellectual disabilities? *Intellectual and Developmental Disabilities, 46*(1), 12–26.

Neisworth, J., & Wolfe, P. S. (2005). *The autism encyclopedia*. Baltimore, MD: Brookes.

Nelson, J. R., Benner, G. J., & Mooney, P. (2008). *Instructional practices for students with behavioral disorders*. New York: Guilford.

Nicholson, L. (2008). *Fetal alcohol syndrome*. Retrieved 3 September 2008, from http://kidshealth.org/parent/medical/brain/fas.html

Norwich, B. (2008). What future for special schools? Conceptual and professional perspectives. *British Journal of Special Education, 35*(3), 136–143.

Nulman, I., Ickowicz, A., Koren, G., & Knittel-Keren, D. (2007). Fetal alcohol spectrum disorders. In I. Brown & M. Percy (Eds.), *A comprehensive guide to intellectual and developmental disabilities* (pp. 213–227). Baltimore, MD: Brookes.

Oberklaid, F., & Kaminsky, L. (2006). *Your child's health* (4th ed.). Prahran, Victoria: Hardie Grant Books.

O'Brien, C. (2005). Modifying learning strategies for classroom success, *Teaching Exceptional Children Plus, 1*(3), (n.p). Retrieved 19 March 2008, from http://escholarship.bc.edu/education/tecplus/vol1/iss3/3

Oddy, M. (2003). Psychosocial consequences of brain injury. In R. J. Greenwood, M. P. Barnes, T. M. McMillan & C. D. Ward (Eds.), *Handbook of neurological rehabilitation* (2nd ed., pp. 453–462). Hove: Psychology Press.

O'Leary, C. (2004). Fetal alcohol syndrome. *Journal of Pediatric Child Health, 40*, 2–7.

Ormrod, J. (2008). *Educational psychology: Developing learners* (6th ed.).Upper Saddle River, NJ: Merrill.

Orsini-Jones, M. (2009). Measures for inclusion: Coping with the challenge of visual impairment and blindness in university undergraduate courses. *Support for learning, 24*(1), 27–34.

Osterhaus, S. A. (2008). *Teaching maths to visually impaired students*. Retrieved 15 October 2008, from http://www.tsbvi.edu/math/teaching.htm

Panda, P. (2001). *Teaching the mentally challenged*. New Delhi: Rajat Publications.

Paul, R. (2007). *Language disorders from infancy through adolescence* (3rd ed.). St Louis, MS: Mosby-Elsevier.

Pellegrino, L. (2007). Cerebral palsy. In M. L. Batshaw, L. Pellegrino & N. J. Roizen (Eds.), *Children with disabilities* (6th ed., pp. 387–408). Baltimore, MD: Brookes.

Peterson, L., Reach, K., & Grabe, S. (2003). Health-related disorders. In E. J. Mash & R. A. Barkley (Eds.), *Child psychopathology* (2nd ed., pp. 716–749). New York: Guilford Press.

Piaget, J. (1952). *The origins of intelligence in children*. New York: International Universities Press.

Pickering, S. J., & Gathercole, S. E. (2004). Distinctive working memory profiles in children with special educational needs. *Educational Psychology, 24*(3), 393–408.

Polloway, E. A., Patton, J. R., & Serna, L. (2008). *Strategies for teaching learners with special needs* (9th ed.). Upper Saddle River, NJ: Pearson-Merrill-Prentice Hall.

Porter, L. (1996). *Student behaviour: Theory and practice for teachers*. Sydney: Allen & Unwin.

Prater, M. A. (2007). *Teaching strategies for students with mild to moderate disabilities*. Boston: Allyn & Bacon.

Prelock, P. A. (2006). *Autism spectrum disorders: Issues in assessment and intervention*. Austin, TX: ProEd.

Pressley, M., & McCormick, C. B. (1995). *Advanced educational psychology for educators, researchers and policymakers*. New York: Harper-Collins.

Prizant, B. M., Wetherby, A. M., Rubin, E., Laurent, A., & Rydell, P. (2005). *The SCERTS Manual: A comprehensive educational approach for children with autism spectrum disorders*. Baltimore: Brookes.

Quirmbach, L. M., Lincoln, A. J., Feinberg-Gizzo, M. J., Ingersoll, B. R., & Andrews, S. M. (2009). Social Stories: Mechanisms of effectiveness in increasing game play in children diagnosed with autism spectrum disorders. *Journal of Autism and Developmental Disorders, 39*(2), 299–321.

Raymond, E. B. (2004). *Learners with mild disabilities: A characteristics approach* (2nd ed.). Boston: Pearson-Allyn & Bacon.

Reddy, L. A., & Pfeiffer, S. I. (2007). Behavioural and emotional symptoms of children and adolescents with Prader-Willi syndrome. *Journal of Autism and Developmental Disorders, 37*(5), 830–839.

Reynolds, C. R. (2000). Sensory integrative therapy. In C. R. Reynolds & E. Fletcher-Janzen (Eds.), *Encyclopedia of special education* (2nd ed., pp. 1622–1623). New York: Wiley.

Roch, M., & Levorato, M. C. (2009). Simple view of reading in Down syndrome: The role of language comprehension and reading skills. *International Journal of Language and Communication Disorders, 44*(2), 206–223.

Roeser, R. J., & Downs, M. P. (2004). *Auditory disorders in school children* (4th ed.). New York: Thieme.

Roizen, N. J. (2007). Down syndrome. In M. L. Batshaw, L. Pellegrino & N. J. Roizen (Eds.), *Children with disabilities* (6th ed., pp. 263–273). Baltimore, MD: Brookes.

Roman, M. A. (1998). The syndrome of Nonverbal Learning Disabilities: Clinical description and applied aspects. *Current Issues in Education, 1,* 1, n.p.

Rubin, K. H., Burgess, K. B., Kennedy, A. E., & Stewart, S. L. (2003). Social withdrawal in childhood. In E. J.Mash & R. A. Barkley (Eds.), *Child psychopathology* (2nd ed., pp. 372–406). New York: Guilford Press.

Runswick-Cole, K. (2008). Between a rock and a hard place: Parents' attitudes to the inclusion of children with special educational needs in mainstream and special schools. *British Journal of Special Education, 35*(3), 173–180.

Salisbury, R. (Ed.). (2008). *Teaching pupils with visual impairment.* Abingdon: Routledge.

Samuels, C. A. (2008). Braille makes a comeback. *Education Week, 27*(43), 27–29.

Santrock, J. W. (2006). *Life-span development* (10th ed.). New York: McGraw Hill.

Scheffler, R. M., Brown, T. T., Fulton, B. D., Hinshaw, S. P., Levine, P., & Stone, S. (2009). Positive association between attention deficit-hyperactivity disorder medication use and academic achievement during elementary school. *Pediatrics, 123*(5), 1273–1279.

Scheuermann, B., & Webber, J. (2002). *Autism: Teaching does make a difference.* Belmont, CA: Wadsworth-Thomson.

Schickedanz, J. A. (1999). *Much more than the ABCs.* Washington, DC: National Association for the Education of Young Children.

Semel, E., & Rosner, S. R. (2003). *Understanding Williams syndrome.* Mahwah, NJ: Erlbaum.

Shane, H. C., & Weiss-Kapp, S. (2008). *Visual language in autism.* San Diego, CA: Plural Publishing.

Simpson, R. L. (2005). Evidence-based practices and students with autism spectrum disorders. *Focus on Autism and Other Developmental Disabilities, 20*(3), 140–49.

Siperstein, G. N., & Rickards, E. P. (2004). *Promoting social success: A curriculum for children with special needs.* Baltimore, MD: Brookes.

Skinner, B. F. (1957). *Verbal behaviour.* New York: Appelton-Century-Crofts.

Smith, D. D. (2007). *Introduction to special education* (6th ed.). Boston: Pearson-Allyn & Bacon.

Smith, T. E. C., Polloway, E. A., Patton, J. R., & Dowdy, C. (2008). *Teaching students with special needs in inclusive settings* (5th ed.). Boston: Pearson-Allyn & Bacon.

Snell, M. E., & Brown, F. (Eds.). (2006). *Instruction of students with severe disabilities* (6th ed.). Upper Saddle River, NJ: Pearson-Merrill-Prentice Hall.

Speech Pathology Australia. (2006). *Students with severe language disorder in Victorian government schools.* Melbourne: SPA, Victorian Branch.

Spohrer, K. (2008). *The teaching assistant's guide to emotional and behavioural difficulties.* London: Continuum.

Stahl, K. A. D., & McKenna, M. C. (2006). *Reading research at work: Foundations of effective practice.* New York: Guilford Press.

Sternberg, R. J. (2003). *Cognitive psychology* (3rd ed.). Florence, KY: Wadsworth.

Stuart, S. (2007). Communication disorders. In M. L. Batshaw, L. Pellegrino & N. J. Roizen (Eds.), *Children with disabilities* (6th ed., pp. 313–323). Baltimore, MD: Brookes.

Sullivan, K. (2004). *The parent's guide to natural health care for children.* Boston: Shambhala.

Tang, N., & Westwood, P. (2007). Worry, general self-efficacy, and school achievement in secondary school students: A perspective from Hong Kong. *Australian Journal of Guidance and Counselling, 17*(1), 68–80.

Tate, R., Smeeth, L., Evans, J., Fletcher, A., Owen, C., & Wolfson, A. R. (2006). *The prevalence of visual impairment in the UK.* London: Royal National Institute of the Blind.

Taub, E., Ramey, S. L., & DeLuca, S. (2004). Efficacy of constraint-induced movement therapy for children with cerebral palsy with asymmetric motor impairment. *Pediatrics, 113*(2), 305–312.

Thompson, S. (1996). *Nonverbal learning disorders.* Retrieved 17 June 2009, from http://www.udel.edu/bkirby/asperger/NLD_SueThompson.html

Tiegerman-Farber, E., & Radziewicz, C. (2008). *Language disorders in children.* Upper Saddle River, NJ: Pearson-Merrill-Prentice Hall.

Topping, K., & Maloney, S. (2005). *The Routledge-Falmer reader in inclusive education.* Abingdon: Routledge-Falmer.

Trezek, B. J., & Malmgren, K. W. (2005). The efficacy of utilizing a phonics treatment package with middle school deaf and hard-of-hearing students. *Journal of Deaf Studies and Deaf Education, 10*(3), 256–271.

Turnbull, A., Turnbull, R., & Wehmeyer, M. L. (2007). *Exceptional lives: Special education in today's schools* (5th ed.). Upper Saddle River, NJ: Pearson-Merrill-Prentice Hall.

Vaughn, S., Bos, C., & Schumm, J. S. (2007). *Teaching students who are exceptional, diverse and at risk* (4th ed.). Boston: Allyn & Bacon.

Vieillevoye, S., & Grosbois, N. (2008). Self-regulation during pretend play in children with intellectual disability and in normally developing children. *Research in Developmental Disabilities: A Multidisciplinary Journal, 29*(3), 256–272.

Vukovic, R. K., & Siegel, L. S. (2006). The double-deficit hypothesis: A comprehensive analysis of the evidence. *Journal of Learning Disabilities, 39*(1), 25–47.

Waterhouse, S. (2000). *A positive approach to autism.* London: Jessica Kingsley.

Webber, J., & Plotts, C. A. (2008). *Emotional and behavioural disorders: Theory and practice* (5th ed.). Boston: Pearson-Allyn & Bacon.

Weiten, W. (2001). *Psychology: Themes and variations* (5th ed.). Belmont, CA: Wadsworth-Thomson.

Westwood, P. (2007). *Commonsense methods for children with special educational needs* (5th ed.). Abingdon: Routledge.

Westwood, P. (2008). *What teachers need to know about reading and writing difficulties.* Melbourne: Australian Council for Educational Research.

Wicks-Nelson, R., & Israel, A. C. (2003). *Behavior disorders of childhood* (5th ed.). Upper Saddle River, NJ: Prentice Hall.

Williams, K. R., & Wishart, J. G. (2003). The Son-Rise Program intervention for autism: An investigation into family experiences. *Journal of Intellectual Disability Research, 47*(4/5), 291–299.

Wilson, G. T., Becker, C. B., & Heffernan, K. (2003). Eating disorders. In E. J. Mash & R. A. Barkley (Eds.), *Child psychopathology* (2nd ed., pp. 687–715). New York: Guilford Press.

Wishart, J. (2005). Children with Down's syndrome. In A. Lewis & B. Norwich (Eds.), *Special teaching for special children? Pedagogies for inclusion* (pp. 81–95). Maidenhead: Open University Press.

Wong, Y. Y., & Westwood, P. (2002). The teaching and management of children with autism. *Hong Kong Special Education Forum, 5*(1), 46–72.

Woodcock, K., Oliver, C., & Humphreys, G. (2009). Associations between repetitive questioning, resistance to change, temper outbursts and anxiety in Prader-Willi and fragile X syndromes. *Journal of Intellectual Disability Research, 53*(3), 265–278.

Woodward, B., & Hogenboom, M. (2000). *Autism: A holistic approach*. Edinburgh: Floris Books.

Woodyatt, G., & Sigafoos, J. (1999). Effects of amount and type of social interaction or activity on stereotyped hand mannerisms in individuals with Rett Syndrome. *Australasian Journal of Special Education, 23*(1), 15–24.

Workinger, M. S. (2005). *Cerebral palsy: Resource guide for speech-language pathologists*. Clifton Park, NY: Thomson-Delmar.

Worthington, L. A., & Gargiulo, R. M. (2006) Persons with emotional or behavioral disorders. In R. M. Gargiulo (Ed.), *Special education in contemporary society* (2nd ed., pp. 289–339). Belmont, CA: Wadsworth.

Wright, C. (2006). ADHD in the classroom. *Special Education Perspectives, 15*(2), 3–8.

Yuen, M. T., Westwood, P., & Wong, G. (2008). Self-efficacy perceptions of Chinese primary-age students with specific learning difficulties: A perspective from Hong Kong. *International Journal of Special Education, 23*(2), 110–119.

Zeamon, D., & House, B. (1963). The role of attention in retardate discrimination learning. In N.R. Ellis (Ed.), *Handbook of mental deficiency* (pp. 159–223). New York: McGraw Hill.

Index

Main entries in **bold**

What Teachers Need to Know About ...

The *What Teachers Need to Know About* series aims to refresh and expand basic teaching knowledge and classroom experience. Books in the series provide essential information about a range of subjects necessary for today's teachers to do their jobs effectively. These books are short, easy-to-use guides to the fundamentals of a subject with clear reference to other, more comprehensive, sources of information.

Other titles available in this series:

I Learning Difficulties	I Reading and Writing Difficulties
I Numeracy	I Spelling
I Personal Wellbeing	I Teaching Methods

To order **What Teachers Need to Know About ...**

Visit <http://shop.acer.edu.au>

A Parent's Guide to Learning Difficulties
How to help your child

Peter Westwood
ACER Press, 2008

A Parent's Guide to Learning Difficulties provides parents with a clear explanation of the many causes of children's problems in learning, and contains jargon-free and practical advice for helping children with reading, writing and mathematics. It also explains how previously proven and effective methods can be implemented in home-tutoring situations, as well as in school.

While the focus is on ordinary children with general learning difficulties, the book also provides important information about teaching and managing children with intellectual, physical and sensory disabilities, as well as autism.

A Parent's Guide to Learning Difficulties is full of links to some great online information resources and references to books that you can use to help your child learn.

About the author

Peter Westwood has been a schoolteacher and university lecturer for many years and has taught all age groups. He has published widely on educational subjects. His research and writing focus on students with learning difficulties and special educational needs. He holds awards for excellence in teaching from Flinders University in South Australia and from the University of Hong Kong.

To order **A Parent's Guide to Learning Difficulties**

Visit <http://shop.acer.edu.au>

Other titles from Peter Westwood

Visit <http://www.acer.edu.au/westwood>